Feedback Controllers for the

Other McGraw-Hill Chemical Engineering Books of Interest

BASTA • *Shreve's Chemical Process Industries Handbook*
BRUNNER • *Hazardous Waste Incineration*
COOK, DUMONT • *Process Drying Practice*
CHOPEY • *Handbook of Chemical Engineering Calculations*
CHOPEY • *Environmental Engineering in the Process Plant*
DEAN • *Lange's Handbook of Chemistry*
DILLON • *Materials Selection for the Chemical Process Industries*
FREEMAN • *Hazardous Waste Minimization*
FREEMAN • *Standard Handbook of Hazardous Waste Treatment and Disposal*
GRANT, GRANT • *Grant & Hackh's Chemical Dictionary*
KISTER • *Distillation Operation*
KISTER • *Distillation Design*
KOLLURU • *Environmental Strategies Handbook*
LEVIN, GEALT • *Biotreatment of Industrial Hazardous Waste*
MANSFIELD • *Engineering Design for Process Facilities*
MCGEE • *Molecular Engineering*
MILLER • *Flow Measurement Handbook*
PALLUZI • *Pilot Plant Design, Construction and Operation*
PERRY, GREEN • *Perry's Chemical Engineers' Handbook*
POWER • *Steam Jet Ejectors for the Process Industries*
REID ET AL. • *Properties of Gases and Liquids*
REIST • *Introduction to Aerosol Science*
SANDLER, LUCKIEWICZ • *Practical Process Engineering*
SATTERFIELD • *Heterogeneous Catalysis in Practice*
SHINSKEY • *Process Control Systems*
SHUGAR, DEAN • *The Chemist's Ready Reference Handbook*
SHUGAR, BALLINGER • *The Chemical Technicians' Ready Reference Handbook*
SMITH, VAN LAAN • *Piping and Pipe Support Systems*
STOCK • *AI in Process Control*
TATTERSON • *Fluid Mixing and Gas Dispersion in Agitate Tanks*
TATTERSON • *Scale-Up of Industrial Mixing Processes*
WILLIG • *Environmental TQM*
YOKELL • *A Working Guide to Shell-and-Tube Heat Exchangers*

Feedback Controllers for the Process Industries

F. G. Shinskey
The Foxboro Company

McGraw-Hill, Inc.
New York San Francisco Washington, D.C. Auckland Bogotá
Caracas Lisbon London Madrid Mexico City Milan
Montreal New Delhi San Juan Singapore
Sydney Tokyo Toronto

Library of Congress Cataloging-in-Publication Data

Shinskey, F. Greg.
Feedback controllers for the process industries / F. Greg Shinskey.
p. cm.
Includes index.
ISBN 0-07-056905-3
1. Feedback control systems. I. Title.
TJ216.S46 1994
629.8'3—dc20 94-2745
CIP

1 2 3 4 5 6 7 8 9 0 DOC/DOC 9 0 9 8 7 6 5 4

ISBN 0-07-056905-3

The sponsoring editor for this book was Gail F. Nalven, the editing supervisor was Kimberly A. Goff, and the production supervisor was Donald F. Schmidt. This book was set in Century Schoolbook by McGraw-Hill's Professional Book Group composition unit.

Printed and bound by R. R. Donnelley & Sons Company.

Information contained in this work has been obtained by McGraw-Hill, Inc., from sources believed to be reliable. However, neither McGraw-Hill nor its authors guarantees the accuracy or completeness of any information published herein, and neither McGraw-Hill nor its authors shall be responsible for any errors, omissions, or damages arising out of use of this information. This work is published with the understanding that McGraw-Hill and its authors are supplying information, but are not attempting to render engineering or other professional services. If such services are required, the assistance of an appropriate professional should be sought.

Contents

Preface xi

Part 1. Performance Objectives

Chapter 1. Performance Criteria 3

Limits of Safe Operation 4
Independence of Controls and Safety Interlocks 4
Cost of Automatic Shutdown 5
Error Magnitude and Plant Efficiency 6
Meeting Quality Specifications 7
Quality Giveaway 8
Energy Use, Capacity Limits 9
Integrated Error Functions 10
Integrated Error (IE) 10
Integrated Absolute Error (IAE) 11
Integrated Square Error (ISE) 12
Integral of Time and Absolute Error (ITAE) 12
Output Weighting Functions 12
Averaging Level and Pressure Control 13
Minimum-Time versus Minimum-Energy Operation 14
Statistical Error Functions 14
Exponentially Weighted Moving Average 15
Quality Distribution Curves 16
Automatic Set-Point Biasing 18
Notation 19
References 19

Chapter 2. Theoretical Limits of Performance 21

Load Characteristics 22
Point of Entry 22
Disturbance Shape 24
Load Response Curves 25
Pure Deadtime 26
Non-Self-Regulating Processes 27

Positive Self-Regulation 29
Negative Self-Regulation 31
Effect of a Secondary Lag 32
Multiple Noninteracting Lags 33
Multiple Interacting Lags 34
Distributed Lags 37
Set-Point Response Curves 38
First-Order Processes 39
Effects of Secondary Lags 40
Performance and Robustness Measures 42
Control-Loop Robustness 42
Notation 44
References 45

Part 2. Linear Controllers

Chapter 3. PID Controllers 49

Proportional Control 49
Bias, Offset, Manual Reset 50
Zero-Load Processes 51
Averaging-Level Control 52
Proportional plus Derivative Control 55
Derivative of the Controlled Variable 56
Derivative Filtering 57
Two-Capacity Processes 58
Integral Control 61
Elimination of Offset 62
Phase and Gain 63
Proportional plus Integral Controllers 65
Phase and Gain 66
Integrated Error 66
Controller Structure 67
Integral Windup 68
PID Controllers 68
Two-Stage Interacting Controllers 70
Single-Stage Interacting Controllers 71
Performance and Robustness Comparisons 73
Notation 75
References 76

Chapter 4. Sampling Effects 77

Controlling Sampled Variables 77
Phase Lag of Sampling 78
Digital Filtering 81
Controlling Sample-Dominant Processes 83
Control of Sampled Lag-Dominant Processes 84
Synchronized Sampling Control 87
Digital PID Controllers 88
Incremental and Positional Algorithms 89
Derivative Gain Limit 92

Sampled Integral Controller 93
Uncertainty in Load Responses 94
Resolution and Threshold Problems 96
Limit Cycling 96
Offset 97
Derivative Spiking 98
Notation 99
References 100

Chapter 5. Model-Based Controllers 101

Internal Model Control (IMC) 102
Pole Cancellation 102
Set-Point Response 104
Load Response 104
The Analytical Predictor 107
Direct Synthesis Controllers 109
The Dahlin Controller 109
Second-Order Controllers 112
Model Predictive Control (MPC) 113
The Smith Predictor 114
Matrix-Based Controllers 116
The PIDτ_d Controller 119
Controller Structure 120
Phase and Gain 122
Optimal Filtering 124
Performance of Model-Based Controllers 125
First-Order Processes 126
Higher-Order Processes 128
Effect of the Sample Interval 129
Robustness of Model-Based Controllers 131
Gain Margins 131
Deadtime Margin 132
Robustness versus Performance 136
Notation 138
References 139

Part 3. Controller Tuning

Chapter 6. Manual Tuning Methods 143

Open-Loop Process Identification 144
Sinusoidal Testing 144
Step Responses 145
A Single Pulse 148
A Doublet Pulse 150
Pseudorandom Binary Sequence 151
Closed-Loop Process Identification 152
Proportional Cycling 153
Limit Cycling 155
Open-Loop Tuning Rules 157
First- and Second-Order Processes 158

Multi-Lag Processes 161
Adjustments for Sampling and Filtering 163
Adjustment for Load Lag 164
Closed-Loop Tuning Rules 166
Initial Settings 166
Fine Tuning 167
Tuning for Set-Point Changes 172
The Integrated-Error Problem 173
Set-Point Filters 176
Effects of Saturation 178
Tuning for Robustness 179
The Deadtime Process 179
The Non-Self-Regulating Process 180
Notation 182
References 183

Chapter 7. Self-Tuning Controllers 185

Pretuning 186
Estimating Noise Level 187
Deadband and Inverse Response 189
Step or Pulse Test? 190
Relay Cycling 191
Performance-Feedback Adaptation 192
Pattern Recognition 193
Covergence Toward the Targets 195
Sinusoidal Disturbances 198
Monitoring the Output 200
Effects of Nonlinearities 201
Output Characterization 201
Input Characterization 202
Adaptive pH Control 206
Batch Processes 207
Notation 209
References 210

Part 4. Nonlinear Elements

Chapter 8. Nonlinear Controllers 213

Multiple-State Controllers 214
Two-State Controllers 214
Three-State Control 216
Pulse-Duration Control 218
Adding Proportional and Derivative or Integral Action 220
Dual-Mode Control 223
Time-Optimal Control 223
Application to Batch Reactors 225
Multiple-State Inputs 226
Using Models 228
Using Fuzzy Logic 228
Nonlinear PID Controllers 231

Error-Squared Controllers 231
Segmental Nonlinear Control 233
Output Characterization 234
Notation 237
References 238

Chapter 9. Constrained Operation 239

Auto-Manual Transfer 240
Bumps During Transfer 240
Achieving True Bumpless Transfer 242
Bumpless Transfer with Proportional Control 244
Power Interruptions 244
Output Limits 245
Position Limits 245
The Output Characterizer 247
Velocity Limits 248
The Tracking Mode 250
Output Tracking 250
Set-Point Tracking 251
Integral Windup Protection 251
Stop Integration 252
The Batch Switch 254
Back-Calculation 259
Multiloop Systems 259
Cascade Control 260
Adding Feedforward Control 262
Decoupling Systems 266
Selective Control Loops 268
Notation 272
References 272

Index 273

Preface

Much confusion currently exists about the performance of feedback controllers relative to one another and relative to what is even possible to expect. For example, model-predictive controllers are being sold for regulation where a proportional-integral-derivative (PID) controller could do better, and PID controllers are being viewed as outdated or unworthy of use in any modern control system. The reason for much of this confusion is the multifaceted nature of process control. Some, familiar with one or two facets, may feel that they have a universal solution to all process-control problems, when, in reality, their experience or perspective may not fit many or possibly most field applications. Academics are particularly susceptible to the narrow view in that most of their time is spent in a clean mathematical environment and a misapplication has no direct impact. Some even refuse to admit the existence of deadtime, although it can be observed wherever materials are transported. This book is an attempt to describe what, in fact, works and what does not in a plant situation, but based on very solid theory.

To be sure, process control sprung from humble beginnings, when mechanical levers and pneumatic bellows did most of the controlling and tinkerers such as John Ziegler and Nate Nichols worked out tuning rules using circular-chart records. As well known as the Ziegler-Nichols tuning rules seem to be, few realize that they apply strictly to step-load changes at the controller output on lag-dominant processes using the Taylor Fulscope controller. Try them on another process or with another controller or expect them to hold up for a load pulse or step at the process output or a set-point change, and you may be disappointed. Yet the fact that they have stood the test of 50 years of use bears witness to their fundamental worth and underlying truth. That they fail to meet your particular expectations in a given application does not mean that they are wrong or worthless. And while empirically derived, they do have a firm theoretical basis (which their authors may have known, but I only discovered years later).

Process-control theory has developed along several parallel paths. The frequency-response method used with electronic devices during World War II was applied to controllers and fluid processes. While this technology shed light on controllers, since they were mechanical and electronic devices, it was not very useful when applied to the processes being controlled. They were too slow and nonlinear to yield much information from frequency testing. While this method was then rarely used in the process industries, it continued to be taught in universities. Time-domain analysis is more applicable to the processes themselves—and easier to learn as well.

Later came optimal control theory and state-space analysis. Although applicable to aircraft and space vehicles, these approaches did not suit process control particularly well and were not adopted by industry. Yet they were taught extensively in universities, with the result that graduating students had to learn process control over again upon entering a plant environment.

The most recent trend in schools is toward model predictive and internal model control. Being based on a dynamic model of the process being controlled, they seem to have the requisite characteristics to succeed in a plant environment. If the process contains deadtime and nonlinearities, the model will include them. Control theory seems at last to be making a penetration into the plant environment.

Yet one unresolved issue remains—tuning. Proponents of model-based control hoped to have avoided tuning, and with good reason. Ziegler-Nichols rules were complicated enough for most practitioners, even in their limited scope. And the observation that most PID controllers have their derivative term set to zero indicates that many degrees of freedom only serve to confuse most people. The models being used in Smith predictors have 3 parameters to set in addition to the controller, and matrix-type controllers have as many as 30 parameters. How can all these be tuned?

In fact, they are not intended to be tuned at all. The model is matched to the process as closely as its complexity and testing accuracy will permit, and then it is held there until performance degradation indicates that another test is required. The operator is then given one adjustment over controller response—essentially the time constant of a filter that determines how fast the controller moves its output. As will be demonstrated, this is not enough to produce acceptable load-response performance on lag-dominant processes.

A look at the history of "one-knob" tuning does not impress. Taylor Instrument Companies produced a one-knob Bi-Act controller, and The Foxboro Company released its Model 59 controller in the 1950s; both were proportional plus integral controllers intended for flow control, and both were short-lived. Their knobs were uncalibrated or had no relevance to the process characteristics, and the controllers were too inflexible.

However, the history of model-predictive control is also barren. Otto J. M. Smith disclosed the Smith predictor in 1957,[1] but it was not even mentioned in Liptak's *Instrument Engineer's Handbook*[2] of 1970 or 1985. To a certain extent, this is understandable—the predictor requires a deadtime simulator, which was not commonly available before the advent of digital controllers. In addition, however, the need to set three model parameters and two controller modes certainly had to discourage users. Furthermore, its performance improvement exacted another price—reduced robustness. It was principally the work of E. B. Dahlin,[3] applying his model-predictive controller to paper machines, that brought attention to the method. His application benefited from two considerations:

1. Model-predictive control is most effective on deadtime-dominant processes (such as paper machines).
2. Stability requires model and process deadtimes to be matched (achievable by measuring machine speed).

The first condition provides higher performance than available with PID control, and the second provides the robustness that model-predictive controllers generally lack.

While not arguing with this success, the recent widespread application of model-predictive controllers to lag-dominant processes such as distillation columns is ill-advised. As will be demonstrated, performance declines exponentially with the ratio of lag-time constant to deadtime unless the model is intentionally *mismatched* against the process to maximize performance. This amounts to *tuning*, however. Because this is inconsistent with presently accepted practice and involves a combination of skill and empirical work, it is not promoted or even considered by proponents of model-based control.

My investigation into high-performance control exposed its dangerous cutting edge: As controller performance approaches 100 percent (of best possible), its robustness approaches zero. In other words, high-performance control teeters on the brink of instability. The high-performance controller is difficult to tune, demanding accuracy in its settings, and is extremely sensitive to parametric variations in the process being controlled. The price of performance comes high. This alone is enough to explain the staying power of the PI controller—it is extremely robust, although of relatively low performance.

Unfortunately, low performance does not guarantee robustness. A model-based controller matched to a lag-dominant process can have both poor performance and poor robustness at the same time. And if a filter or slow sampling or gain reduction is used to improve robustness, performance suffers even more.

The emphasis is on controller performance in this book, by placing it up front, in Part I. Chapter 1 examines economic measures of performance by describing the role of process control in plant economics. Chapter 2 then defines the theoretical limits of feedback-controller performance to set realistic goals for both the controller and the process. This theoretical limit is then the benchmark for performance evaluation of all types of controllers in the chapters that follow.

Part II introduces linear controllers, beginning with PID and its component parts in Chapter 3. This is followed by a presentation of sampling, a necessary evil in digital controllers. Chapter 5 then concentrates on several types of model-based controllers, pointing out their similarities, performance advantages, and limitations. An outgrowth of this study is the hybrid PIDτ_d controller, which combines high-performance with tunability.

Part III concentrates on controller tuning, first manually in Chapter 6, where the principles of dynamic modeling and performance optimization are developed. The procedures are then automated in self-tuning controllers in Chapter 7. Part IV investigates nonlinear elements, first by presenting various nonlinear controllers in Chapter 8 and finally by examining the nonlinear operating regions of constrained linear controllers in Chapter 9.

In my previous books, dynamic analysis has been confined to the time domain, because this is familiar to practitioners and easily assimilated by novices. However, certain particular aspects of controllers are easier to examine and compare by using frequency-response analysis and transfer functions. While the use of these methods will be minimized, they do assume on the part of the reader a grasp of process-control fundamentals and operational calculus. The control theory presented here is rigorous without being complex and is demonstrated by numerous simulations.

F. G. Shinskey

References

1. Smith, O. J. M., "Close Control of Loops with Dead Time," *Chem. Eng. Prog.* 53(5): 217–219, May 1957.
2. Lipták, B. G., *Instrument Engineers' Handbook*, Vol. II, Chilton, Philadelphia, 1970; revised edition, 1985.
3. Dahlin, E. B., "Designing and Tuning Digital Controllers," *Instr. Contr. Sys.* 41(6): June, 1968.

Feedback Controllers for the Process Industries

Part

1

Performance Objectives

Chapter

1

Performance Criteria

The purpose of a controller is to keep a controlled variable at its desired value in the presence of disturbances from various sources and to cause it to follow changes in said desired value as closely as possible. The former—that of maintaining constancy in the presence of disturbances—is called *regulation,* while the latter—that of following changes in the desired value—is termed *servo response.* In mechanical systems such as machines and vehicles, servo response is the primary consideration, disturbances being relatively minor and intermittent. In control of fluid processes, however, regulation is the more important and more difficult function, in that unmeasured disturbances are frequent and severe; changes in the desired value (set point) tend to be common only in batch as opposed to continuous processes and in secondary, or "slave," loops. These distinctions are expanded and examined in more detail as individual applications are presented.

This introductory chapter is intended to establish the relationship between the ability of a controller to approach the preceding goals and the economic penalties for failing to do so. If a controller serves no economic function, then it has no justification in today's workplace. Be assured that safety and environmental protection fall under the economic umbrella, because accidents and pollution exact economic penalties. The issue here is that controllers and their support are costly, and the protection that they provide must justify their expense. Hence return on investment is always the primary consideration in industrial systems, and the controller that provides the earliest economic return represents the best investment. This establishes the need for economic measures of controller performance.

There are several areas where controllers can contribute to the economic performance of the plant being controlled. Each area has its own individual characteristics and needs which the controller must serve. Each is touched on below with respect to its own sources of economic penalties and the role of the controller in mitigating those penalties.

Limits of Safe Operation

Safety is the primary consideration in the operation of any system, be it an appliance, a vehicle, or a process plant. If the system cannot be operated safely, then it will not fulfill its primary productive function dependably or for long enough to repay its investment. The costs of accidents can be excessive—loss of life cannot even be evaluated satisfactorily, and the cost of damage to the environment keeps changing as we learn more about it. These factors therefore cannot really be entered into the economic equation. Safety simply must be built into the operation to minimize the likelihood of an accident through all the foreseen avenues. Control systems can contribute to safe operation, and they should. But they should not be the sole contributors. The plant must first be designed to fail safely, because fail it will, eventually. And the controls must be backed with a completely separate system of interlocks.

Independence of controls and safety interlocks

There should be several layers of protection built into any inherently hazardous operation. For example, a boiler will have several safety valves that lift at a pressure well below the stress limits of the vessel itself. Additionally, there should be a high-pressure interlock that will trip the combustion system before the safety valves lift. Third, there will be a steam-pressure controller that manipulates the firing rate and is intended to keep pressure well below either of the other limits. In this way, the interlocks and safety valves would be exercised only if the pressure-control system failed, which, although unlikely, could happen as a result of a severe upset, operator error, component failure, or some combination thereof.

It is mandatory that the different levels of safety systems have no common mode of failure. For example, the pressure controller and the high-pressure interlock should not use the same pressure transmitter or even the same type of pressure-measuring device. Their power sources should be separate, their signal wiring separately routed, etc., considering all events that might compromise both systems, such as a fire. And of course, both should fail safely, i.e., shut off the source of energy to the process, in the event of loss of either signal or power.

Redundant instrumentation can provide additional protection and should be used for devices whose reliability is lower than others in the rest of the system. Automatic selection should be provided between redundant pairs on a fail-safe basis so that the most likely failure will shut the unit down even if there is no accident. Since this event can be quite costly, a third redundant channel may be justified to protect against a single failure in either direction without causing a false shutdown. This is common practice in controlling the pressure inside balance-draft furnaces; three pressure transmitters send their signals to a median selector which discards the highest and lowest of the three signals. Three comparators are also required to identify errant transmitters. Two-out-of-three logic is also used commonly in protection and control of nuclear power plants.

Safety protection can be excessive to the point where the plant cannot be started or is subject to frequent "nuisance" shutdowns. This encourages operators to find ways around the interlocks, which may expose them to real dangers. Assuming that the controls and interlocks have been designed to be operable, the issue at hand is to avoid shutdowns caused by failure of the controls to keep critical variables from reaching the settings of the safety interlocks.

Cost of automatic shutdown

Loss of production is not the only cost of a shutdown. The shutdown operation itself will waste energy and material stored in the process, which must be removed. And the subsequent startup will require a similar amount of energy and material to be added to reach operating levels again. There also will be a period of time after startup before product quality will be acceptable, further extending production loss and requiring the recycling of off-specification product.

There are other hidden costs as well. Startup always stresses equipment and operators more than continuous production and is a time when most accidents occur.

Another factor is the interconnection that may exist between the tripped unit and others which may depend on it or may share the load with it. A tripped boiler may cause the shutdown of processes using its steam. If several boilers are supplying steam in parallel, production may continue only if others are able to pick up the load lost by the tripped unit. Having enough capacity on-line to continue at the same production rate is probably not economical. However, tripping one of several parallel units also stresses those remaining on-line. If their controls are unable to cope with the sudden load increase posed by the tripped unit, one or more of them may trip as well, which, if no automatic load shedding is in place, could bring down the entire plant. This was the cause of the Northeast power blackout of November of 1965—a component failure in one power station caused an overload which tripped one station after another until the entire

Northeast coast of the United States was in darkness. Decoupling has been added to the power grid to prevent a recurrence, but it could still happen in any individual plant unless similar precautions are taken.

In summary, shutdowns caused by the action of safety interlocks are costly, although their consequences could be far more costly if such interlocks did not operate. However, the cost varies widely from one plant and one situation to another.

Error magnitude and plant efficiency

Some controlled variables have no specific relationship with production efficiency over a range of values. Liquid level is one such example. The precision of level control in a tank or even in a boiler has no economic significance until some specific point is reached where effects other than liquid level appear. The level of feedwater in the drum of a boiler is a classic example. As long as the level is maintained within a range, it has no impact on boiler efficiency. If the level is too high, however, water may be carried out into the superheating tubes, creating stress there, or if the level is too low, evaporating tubes may be uncovered, causing stress there. The selection of the set point for the drum-level controller is typically simply equidistant between two similar evils and the trip points associated with them.

The case where pressure and temperature are controlled is usually different. Danger is associated with excessive values of both, and safety is associated with low values, as is process efficiency. The driving force for production processes is usually created by a pressure and/or temperature difference against the environment. Fluid flows across a pressure difference, and heat flows across a temperature difference.

In the case of a power plant, for example, the efficiency of converting heat into electric power increases with steam pressure and temperature. Their values should therefore be maximized consistent with the strength of the vessels and piping containing the steam. Excessive temperatures can shorten tube life even when transient. Therefore, there exist an optimal operating temperature and pressure that balance operating costs against capital costs, providing a maximum return on investment.

If the plant were always to operate at these optimal conditions, efficiency would be maximized. Because of variations in load, however, which always occur in any production facility, pressure and temperature will vary around their set points to an extent determined by the size of the upsets and the effectiveness of the controls. The upsets may simply have to be accepted as they are, but the controls should be improved to minimize the resulting excursions. Finally, the pressure and temperature set points must be positioned so that the largest expected upset will not cause an excursion that will trip the

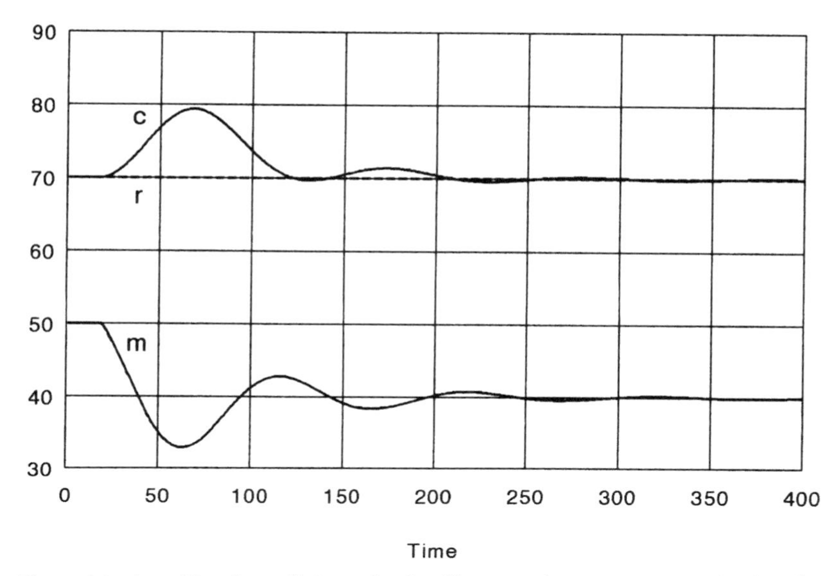

Figure 1.1 A sudden loss of steam load will cause steam pressure c to rise above set point r until the pressure controller is able to reduce firing rate m sufficiently.

unit. Thus the efficiency of the process, as a function of these set points, is also a function of the effectiveness of the controls.

Figure 1.1 describes the type of response observed in boiler steam pressure upon a sudden loss in steam load. Because more steam is still being generated than required for the current load, the balance accumulates as the difference between controlled pressure c and set point r. The pressure controller responds by reducing manipulated firing rate m, but delays and heat capacity in the heat-transfer system continue the generation of excess steam for a while. The pressure rise will be minimized by taking control action as early as possible. Even so, there is a limit to what a feedback controller can do in the presence of said delays—a limit that is carefully defined in the next chapter. Within this limit, a more active controller will reduce the height of the pressure transient and therefore allow the pressure set point to be indexed closer to the optimal efficiency of the unit and the trip point for the safety interlocks. A similar scenario could be recited for steam temperature control.

Meeting Quality Specifications

The quality of materials leaving a production facility must meet certain specifications to be sold or even to be sent to further processing within the same plant. Gasoline, for example, must satisfy the octane label on the pump, as well as vapor-pressure limits set by the government, to name only two specifications. Material in intraplant transfers similarly must meet the requirements of downstream processing such as boiling point, percent solids, water content, etc., depending on the use to which it will be put. Virtually every process therefore needs control over one or more properties of the products that leave it. And much of the revenue lost in the operation of a plant can be attributed to poor quality control.

Quality giveaway

There are two different ways in which quality can be given away by a manufacturer, depending on the direction of a disturbance entering the product-quality loop and the position of the quality set point. Let Fig. 1.1 represent the control of the octane number of a gasoline blend, and note that set point r is positioned precisely at the specified number. An upset in the direction shown in Fig. 1.1 therefore results in octane exceeding the specified number, which is acceptable to the user but costly for the manufacturer. The area under the response curve between controlled variable c and specification r could be expressed in octane-minutes. However, if production continued at a fixed rate expressed in liters per minute, then multiplying that rate by the area under the response curve could represent the octane-liters given away, for which no return was received. (The user pays based on the specification—the label on the pump—not on the actual quality of the product.) The differential in dollars per liter as a function of octane number multiplied by the octane-liters given away yields the dollars given away. Therefore, in this example, the integrated error under the response curve is proportional to dollars given away.

This analysis presumes, however, a linear relationship between cost or value and octane number. While this tends to be true for small variations, the correct nonlinear function may have to be applied to the error between the controlled variable and set point before integration for accuracy on larger upsets.

If the upset happened to be in the other direction with the set point still at the specification limit, product quality would violate the specifications. This event requires precipitous action to be taken, because off-specification material cannot be allowed to leave the plant masquerading as quality product. In this situation, there are several options available, listed below in order of their relative cost:

1. Blend the off-specification product with above-grade product.
2. Upgrade it by reprocessing.
3. Sell it as lower-grade product at a lower price.

4. Burn it as fuel.
5. Dispose of it as waste.

The options are also listed in proportion to the magnitude of the deviation of quality from the specification, although not all may be open for a given product. Still, all the options are costly and to be avoided wherever possible, in that they require a decision to be made. The first action to be taken is the diversion of the off-specification product into alternative storage from which it can later be withdrawn for blending, reprocessing, etc. This allows time for weighing the options and taking action relative to the severity of the problem. However, the simple act of diversion raises a flag on the day's operation, which the operators would very much like to avoid.

Since upsets can be expected in either direction, the set point realistically cannot be positioned at the limit but must be precessed from it by an amount representing the largest deviation likely to occur. The larger this margin, the less likely that a violation will occur. Operators tend to prefer wider margins. However, the margin itself represents product giveaway in the steady state, which may be much higher than that resulting from transient excesses. In this service, a control system that minimizes error magnitude (not integrated error) will minimize giveaway.

If product accumulates downstream in a mixed storage tank, short-lived violations may not be costly, in that current product is continuously blended with that previously made. Only the material leaving the storage tank must meet specifications, and its quality tends to be the average of that made during its residence time in the tank. If the tank has a very long residence time, upward and downward disturbances may be distributed equally, in which case integrated error could approach zero. Then the controller set point could be positioned at the specification limit. However, excessive inventory is costly, too, and decreasingly common.

Energy use, capacity limits

In some cases, there may be no product-giveaway penalty. This would be true if the controlled ingredients in the product had similar values. However, there still may be an energy penalty, in that overpurification does require more energy per unit of product made. This energy penalty exacts a capacity price as well. Most processing equipment is energy-limited; i.e., production capacity is limited by energy transfer rates. With overpurification, the rise in energy use per unit of product made means that less product can be made within the energy limits of the equipment. This cost factor can be extremely important when the product market is not limited.

In some processes, all three of the preceding cost factors are active at the same time. Consider, for example, an evaporator or dryer producing a food product for sale. If the product is overconcentrated or overdried to avoid violating specifications on its allowable water content, solid product is given away in place of water, excessive energy is used to overconcentrate it, and capacity is lost. Moreover, if the overconcentration is extreme, the product could even be discolored or otherwise damaged and the processing equipment fouled. Minimizing quality deviation is especially important here.

Integrated Error Functions

While error magnitude is an important criterion of controller performance, it is also imperative that the error be reduced to zero as quickly as possible in most control loops. The criteria that evaluate this return trajectory are all integrated error functions; however, there are several, giving different weights to error magnitude, duration, and direction. They are presented below with comments relative to their significance and application.

Integrated error (IE)

The *integrated error* is simply the integral of the deviation with respect to time t following a disturbance:

$$\text{IE} = \int e\,dt \tag{1.1}$$

where e is the error or deviation between controlled variable c and set point r.

Integrated error can be calculated analytically as a function of the change in controller output m required to return the error to zero following any disturbance. For the ideal proportional-integral-derivative (PID) controller, the relationship can be derived from the controller algorithm. Consider an initial state of the controller at time t_1 where its output is m_1:

$$m_1 = \frac{100}{P}\left(e + \frac{1}{I}\int_{t_0}^{t_1} e\,dt + D\frac{de}{dt}\right) + m_0 \tag{1.2}$$

where P, I, and D are the settings of the three control modes, and m_0 is the output of the controller at t_0 when it was first placed into automatic operation. If t_1 is identified as a steady state, then e is zero and so is its derivative, leaving the integral term as the only nonzero error term in the equation.

Following a disturbance from this steady state, the controller will eventually restore the error to zero, reaching a new steady state at time t_2. Evaluation of controller Eq. 1.2 at t_2 will again result in only the integral error term being nonzero:

$$m_2 = \frac{100}{P}(\frac{1}{I}\int_{t_0}^{t_2}) + m_0 \tag{1.3}$$

Subtracting the initial from the final state yields the change in controller output required to respond to the disturbance and the integrated error resulting from that change:

$$\Delta m = \frac{100}{PI}\int_{t_1}^{t_2} e\,dt \tag{1.4}$$

Solving for the integrated error yields

$$\int_{t_1}^{t_2} e\,dt = \Delta m\,\frac{PI}{100} \tag{1.5}$$

This will be a very useful expression for evaluating controller performance and indicating how it can be improved. In practice, Eq. 1.5 may require some modification when used with controllers other than the ideal PID. For example, integral term I will be augmented by the sampling interval of a digital controller, by the time constant of any filter inserted in the loop, or by the value of any deadtime compensator used. These specifics will be examined under performance evaluations of the individual controllers.

Integrated absolute error (IAE)

The principal limitation of IE is that it says nothing about the stability of a control loop. Reducing the proportional and integral settings will reduce IE but at the same time also will lower the stability margin. Eventually, settings will be reached that produce a uniform oscillation, but because such oscillation would be distributed uniformly about zero error, there would be no net impact on IE. Consequently, a controller cannot be tuned simply to minimize IE—a stability criterion must be included as well.

All the remaining integrated-error functions contain inherent stability criteria, which is why they are used more commonly than IE. Each will accumulate without limit in the presence of an undamped oscillation and only stop integrating in a steady state where deviation is zero.

The most useful of the integrated-error functions is integrated absolute error (IAE), which is simply the time integral of the absolute value of the deviation of the controlled variable from set point. If we accept that IE is related to economic performance, then minimizing IAE will honor the same goal while ensuring stable return to set point.

The PID controller used in the example in Fig. 1.1 was tuned for minimum IAE. Observe that following the step-load change, the controlled variable overshoots the set point only slightly on the return trajectory, making IE and IAE nearly identical. Therefore, the practice of minimizing IAE virtually ensures minimizing IE as well while providing a very stable response. Additionally, error magnitude is minimized by the same practice. As a consequence, IAE will be found to be the most generally applicable of all the performance criteria. Even where IE may be the preferred measure of performance, the controller should be tuned to minimize IAE.

Integrated square error (ISE)

This criterion squares the error before integrating it, with the result that the sign of negative errors disappears as in IAE. However, squaring weighs large errors more than small ones. While intended to reduce peak error, minimizing ISE may result in extended settling times, since small errors have little weight.

There seems to be no generic relationship between integrated square error (ISE) and economic performance, so it cannot be recommended as widely as IAE. In fact, its popularity in academic circles seems to be due to its mathematical convenience rather than to anything else. It is more amenable to optimization routines than IAE, which must be minimized by trial and error.

Integral of time and absolute error (ITAE)

This criterion multiplies the absolute value of the error by the time since it began, followed by integration. It has the opposite effect of ISE, tending to penalize long-term errors more than short transients. The minimum ITAE load-response curve will thereby tend to have a higher peak and shorter duration than minimum IAE and ISE curves. This criterion seems even less useful than ISE, in that it tends to allow larger peaks, which is generally less acceptable, in an effort to minimize duration, which seems rather arbitrary.

In fact, the principal incentive for minimizing error duration would apply to batch-process operation, where production can be increased by reducing the time required to change states. In this light, ITAE might be more suitable to evaluate set-point response than load response, since set points are changed frequently in batch operations and each change has a known starting time.

Output Weighting Functions

Much of the work in optimal control theory involves weighting functions applied to the output of the controller as well as to the input.

There could be several reasons for such a function. If the controller manipulated two streams in split range—such as heating and cooling or acid and base or engine and brakes—excessive output motion could result in wasting these resources against each other. Another possible reason to conserve output motion would be to reduce wear on valve actuators and seals.

In most control loops in the fluid-process industries, however, output motion is not an important criterion—in fact, it is rarely given any consideration at all. Tight control usually means rapid manipulation within limits of turndown which the process can accept. One advanced feature that might be valuable in eliminating excessive output motion would be an adaptive output filter that controls the standard deviation of the output at some desired level measured over a designated time span.

Averaging level and pressure control

There is one important exception to the rule, however, where output motion is of more concern than deviation from set point. This case is called *averaging-level* or *averaging-pressure control.* Its purpose is to minimize the rate of change of the manipulated flow leaving a vessel in response to random upsets in its inflow so that disturbances to downstream processes will be minimal. The vessel in question is called a *surge tank,* and its function is to act as a flow filter having a time constant as long as possible. The criterion applied to the controlled variable (level in the case of a liquid or pressure for a gas) is simply to avoid exceeding some arbitrary maximum and minimum limits.

None of the integrated-error criteria apply to this situation, and even error magnitude should not be minimized—if anything, it should be maximized to the preceding limits. As a consequence, none of the control methods used to minimize errors or their integrals qualifies as an averaging controller. This application requires a different controller and a different approach to tuning it. Reference 1 reports on audits of five petroleum refineries where advanced controls have been installed and indicates that "the most commonly mistuned loops in the audited refineries were level controllers, where the operating objective is usually to keep flow steady and accept level swings, but where tuners prefer the opposite." The simplest solution is to use a proportional controller as described in Chap. 3, integral action being counter to the control objective and destabilizing as well. The principal obstacle here is that the objective is opposite to that of most control systems and is not well understood, especially by those who must operate the plant.

Minimum-time versus minimum-energy operation

Usually, no limits are placed on controller outputs, except mechanical limits on valve actuators. In this way, there is no interference with the action the controller takes in attempting to satisfy its objectives for the controlled variable. In a batch process, this would mean heating and cooling at maximal rates so as to minimize the time required to move a process from one steady state to another. Much effort has been put into pursuit of minimum-time control, especially in batch reactors.

To heat a vessel to a controlled temperature in minimum time requires the application of full heat until the temperature has come within a calculated difference below set point, followed by maximum cooling for a predetermined time. Then normal heating or cooling may be applied consistent with the estimated heat load on the vessel in order to hold temperature constant. The closeness of approach to set point at the moment of switching from heating to cooling depends on the strength of the cooling source. The same method could be used in propelling a vehicle from one point to another in minimum time: full speed almost to the destination, followed by full braking, with the switching point determined by the available braking. This method has been used in controlling batch reactors, but with some difficulty, in that the effectiveness of both heating and cooling changes with temperature, requiring a different switching point for every set of conditions.

An alternative strategy is that of minimum-energy control. By making cooling unavailable, a minimum amount of heat will have to be applied to keep from overshooting the temperature set point. Making heating unavailable during conditions requiring cooling serves the same purpose then. This is a conservative strategy from the standpoint of material resources, but it requires more time to complete the operation. An optimal strategy where both time and resources are conserved would fall in between these two, achieved by partially limiting resources.

Statistical Error Functions

Much emphasis is being given lately to the statistical properties of product-quality measurements. This work originated in discrete-parts manufacture and has migrated into process control under the name of *statistical process control.* The objectives are essentially the same whether the products are parts or fluids, but the methods found successful with the former are not necessarily applicable to the latter.

Exponentially weighted moving average

In discrete-parts manufacture, averages and standard deviations are taken of products segregated into lots, each lot representing the production from a given plant unit over a given time. Mean, variance, and standard deviation are calculated for n samples taken from a lot using formulas so well known that they will not be repeated here.

In fluid process control, however, aggregation of products into lots takes place only at the packaging or distribution end of the plant. By this time, quality should have been controlled already, and nothing more can be done with the product. Tank-size lots may be sampled, but their contents represent the accumulation of the product over time. To be sure, individual lots may be compared statistically as discrete parts are, but this is not where control is applied.

Quality control is applied to the product as or before it leaves the process where that quality is determined. In a blending process, for example, product composition is measured and controlled immediately after the components are mixed and before the product reaches its storage tank. In the past, blending was conducted in the tank itself, batchwise, and this may still be practiced for small quantities. But now, most blending is continuous, with continuous control over product quality. This is the practice even when treating wastewater.

The same holds true for distillation. Batch distillation was once quite common, with each cut accumulated in its own tank, and quality measured there. Most distillation is now continuous, however, with tankage between units essentially eliminated; composition is controlled as the product leaves the column.

The mean value of the composition of a flowing product is then not to be calculated as if the product were accumulated as a lot but considering that it is flowing into and out of a tank at the same time. If the tank had no mixing of its contents at all, then product would flow through it as piece parts on a conveyor belt. The mean and variance would then be calculated on the moving contents of the tank, each sample having equal weight and remaining in the tank for an equal length of time. But mixing always takes place in tanks, even in the absence of an agitator, with the result that the product leaving represents the average composition of the contents of the tank. This average composition is expressed as the exponentially weighted moving average.[2]

This measure has the property that the weight given to a particular sample decreases exponentially with time as that sample is blended with those having entered the tank earlier. It is very simply calculated as

$$\overline{x} = \overline{x} + (x_t - \overline{x})/n \tag{1.6}$$

where x_t is the composition of the sample entering the tank, and n is the number of the samples contained in the tank. The latter is the residence time of the fluid in the tank divided by the time interval between samples. Residence time can be expressed as $\tau = V/F$, where V is the volume of fluid contained in the tank, and F is the current flow rate. If the flow into the tank is variable, then n will vary accordingly and should be calculated as a function of the flow measured at the time a sample is taken:

$$n = V/F\Delta t \tag{1.7}$$

where Δt is the interval between samples.

Having calculated the average composition in the tank, the variance is determined as follows:

$$\sigma_x^2 = \sigma_x^2 + [(x_t - \overline{x})^2 - \sigma_x^2]/n \tag{1.8}$$

and then standard deviation σ_x is the square root. While the procedure was explained in the context of a tank having a definite residence time, it may be applied to any flowing stream without a downstream tank, given a residence time or n assigned arbitrarily.

Quality-distribution curves

It is commonly accepted that product quality distribution follows the classic gaussian bell-shaped curve about the mean, as shown in Fig. 1.2.

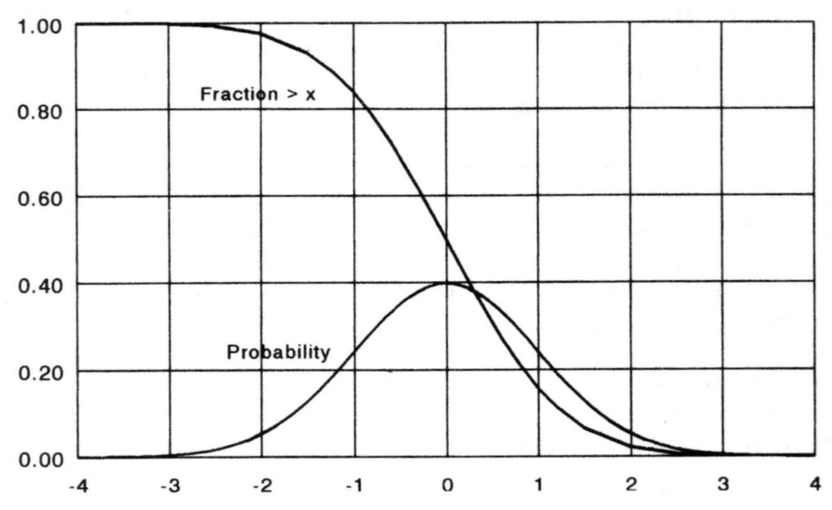

Figure 1.2 Product composition may be distributed uniformly about the mean as in this gaussian distribution curve.

This has been amply demonstrated by statisticians for many population samples. However, it may not be accurate for measurements of the composition of fluid products in several respects. One problem is the existence of a closed loop giving rise to oscillations in the measurement. Another problem may arise when a controlled component approaches zero, which represents an absolute limit on the variable. A third possibility is the presence of uncorrelated, or "white," noise, which may have a flat distribution.

Closed loops tend toward oscillatory behavior, which in a linear system appears as a sine wave. If a loop is oscillating uniformly, the distribution of the controlled variable appears to be bimodal, i.e., having two peaks, as shown in Fig. 1.3. The reason for the low value corresponding to the mean is that the velocity of the sine wave is maximum there; conversely, the velocity is zero at the peaks, maximizing the distribution there. About the only statistical information that can be obtained from such a distribution curve is that the loop is cycling, which is easier to observe from the time record of the controlled variable. Damping in the loop will tend to bring the two peaks closer together, narrowing the distribution as the oscillation eventually disappears.

Products of refining operations such as distillation, evaporation, and drying become progressively harder to purify as their impurity approaches zero. As a result, oscillations in these closed loops can be

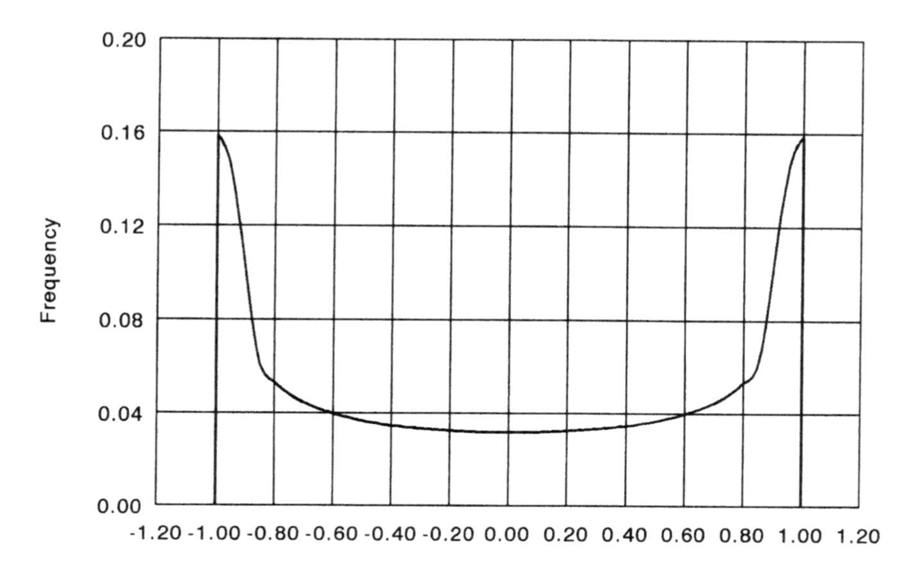

Figure 1.3 Uniform oscillation in a linear closed loop is sinusoidal, producing a bimodal distribution curve.

decidedly nonsinusoidal, having sharp peaks at high impurity levels and flat valleys as zero is approached. This behavior produces a highly skewed distribution curve having a large peak near zero and a long, thin tail reaching to high impurity levels.

Regardless of whether the distribution is uniform, skewed, or bimodal, the mean value of the controlled variable is at the set point if the controller is linear and if enough time has elapsed for increasing and decreasing load changes to balance. If nonlinear compensation is added to the controller to improve response in the presence of the nonlinearity described in the preceding paragraph, there will be an offset between the mean and the set point in proportion to the size of the disturbances entering the loop. Also, if disturbances have been principally in one direction, the mean will not coincide with the set point, which is the case for the response in Fig. 1.1 over a single upset. Any meaningful statistical analysis will have to take this last factor into account.

Automatic set-point biasing

As described earlier under Quality Giveaway, it is common for operators to position the set point of the product-quality controller away from the specification limit on product quality by the expected maximum deviation in the controlled variable. If they have no information on the probable size of disturbances, they will set the margin for the worst case. This, unfortunately, maximizes the economic penalty for that production unit.

A patent[3] by the author describes an automatic means for biasing the set point away from the specification limit consistent with the current standard deviation in the controlled variable. In this way, the margin is kept as low as possible while minimizing the likelihood of violating product specifications.

Superimposed on the bell-shaped distribution curve in Fig. 1.2 is the integral of that curve. It represents the fraction of the population (the area under the distribution curve) having a value greater than x on a scale of σ. If product composition is distributed uniformly about the set point, then positioning the set point 2σ away from the specification limit will, on average, yield product violating specifications only 2.3 percent of the time. This or a similar level of risk may be acceptable, given some blending capacity downstream. Increasing the bias to 3σ reduces the probability of violation to 0.1 percent. The patent describes calculating the standard deviation of the controlled variable over some given time and biasing the set point away from the specification limit in proportion to the standard deviation. In this way, the bias or margin will be as low as the variability of the product will allow.

The patent also describes a method applicable to a skewed distribution, where the relationship between σ and the area under the distribution curve may be unknown. A simple area ratio is applied; the controlled variable is compared to the specification limit in an integrator whose gain on the unacceptable side is several times that of its gain on the acceptable side of the specification limit. The multiplication factor represents the ratio of acceptable product to unacceptable product, such as 50:1. This area ratio will be attained by the integrator adjusting the set point of the composition controller.

Notation

c	Controlled variable
d	Derivative operator
D	Derivative time
e	Deviation from set point
F	Flow rate
I	Integral time
IE	Integrated error
IAE	Integrated absolute error
ISE	Integrated square error
ITAE	Integral of time and absolute error
m	Manipulated variable
n	Number of samples
P	Proportional band
r	Set point
t	Time
Δt	Sample interval
x	Composition
x_t	Composition entering
$\overline{x}$	Mean value of x
σ	Standard deviation
τ	Residence time

References

1. Friedman, Y. Z., "Avoiding Advanced Control Project Mistakes," *Hydrocarbon Proc.*, October 1992, pp. 115–120.
2. Hunter, J. S., "The Exponentially Weighted Moving Average," *J. Quality Tech.* 18(4): 203–210, 1986.
3. Shinskey, F. G., Method and Apparatus for Statistical Set-Point Bias Control, U. S. Patent No. 4,855,897, August 8, 1989.

Chapter

2

Theoretical Limits of Performance

The concept of theoretical limits to feedback-controller performance is relatively new, having been introduced by the author in 1988.[1] It provides a realistic performance goal for controller selection and design and an absolute measure of performance useful for comparing controllers of various types. It also can allow the estimation of peak values of controlled variables resulting from severe upsets to determine the cost or danger incurred thereby. The concept depends on the existence of deadtime in the loop, since in its absence, there is no theoretical performance limit. This is not a real obstacle, however, because many, if not most, processes exhibit measurable deadtime, even in the absence of an identifiable transportation lag. While multicapacity processes may have no real deadtime, an effective deadtime is created by the accumulation of many lags, both to the eye of an observer attempting to characterize their dynamics and to the controller attempting to regulate them.

The real test of controller performance is in response to unmeasured step changes in load. Many engineers work to optimize set-point response instead, which is a much easier task in that the controller is aware of its set point but not the load. However, tuning a controller exclusively for set-point response may seriously compromise its load-response capability. This chapter examines both responses at the limit of performance.

Robustness is a measure of tolerance that a control loop has for variations in its properties before instability is reached. A low measure of robustness means that the controller must be tuned accurately, will be difficult to tune, and is not likely to remain stable in the presence of process variations. It will be shown that as controller performance is pushed to the absolute theoretical limit, robustness axiomatically approaches zero. This is not in itself an indication of maximal performance, however, because robustness can approach zero without performance being maximal—surely an undesirable combination.

Load Characteristics

There are many possible points of entry of a load disturbance into a closed loop, as well as many shapes of disturbance. Both have a heavy impact on the performance of the controller and on its tuning as well. In fact, engineers can come to entirely different conclusions on the effectiveness of a given controller or how to tune it without realizing the role these factors play or making much note of them in their studies. The gap between academic teaching and industrial practice is largely due to disagreement about and misunderstanding of the load characteristics encountered in process plants. The following is intended to clarify these issues.

Point of entry

The general case for the load input can be represented by the block diagram of Fig. 2.1. Here, the load encounters a steady-state gain K_q

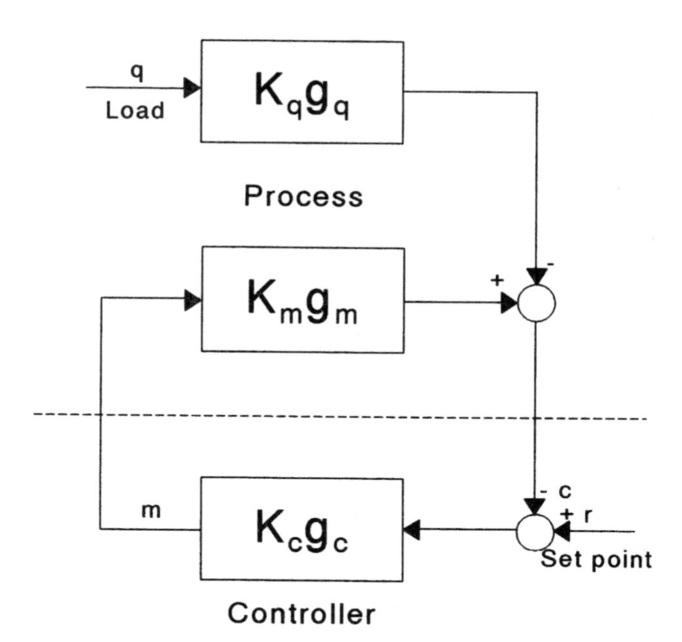

Figure 2.1 The general case has the load variable entering the loop through its own set of gains.

and a dynamic-gain vector $\mathbf{g}_q$ which are different from K_m and $\mathbf{g}_m$ in the path of the controller output. In a feedforward system, the ratio of the steady-state gains is used as the feedforward steady-state compensator, and the ratio of the two dynamic-gain vectors is the required dynamic compensation. Here, however, the issue is the response of the controlled variable and feedback controller to the load.

The dynamic-gain vector includes both a magnitude and a phase angle, which are typically functions of the period or frequency of the signal passing through. The vector is a product of the vectors of all the dynamic elements appearing in the path of the signal and may include deadtime, lags and possibly a lead, and sometimes an integrator; rarely, a negative lead or lag will be encountered. Each of these elements is described later with regard to its contribution to $\mathbf{g}_m$, where it lies within the closed loop.

Steady-state gain K_q is unrelated to the point of entry of the load, being a function only of the steady-state model of the process. Dynamic-gain vector $\mathbf{g}_q$, however, can indicate that point of entry. If $\mathbf{g}_q$ is zero, then the load enters the loop at the process output, affecting the controlled variable immediately. Although this point of entry is seen commonly in academic papers, it is almost never observed in industrial applications. The only disturbance likely to enter the loop at that point is measurement noise, which, by definition, is a transient whose frequency range is too high for the controller to do anything about. While the noise factor is important and may require filtering or prevent the use of derivative action, it is not and should not be considered a load disturbance. If load changes are in fact introduced at this point, the controller will respond instantaneously, which is quite different from the usual response to load changes. This issue will be reexamined under the subject of model-based control, where it severely limits controller performance.

The opposite case is the point of entry at the controller output, such that $\mathbf{g}_q = \mathbf{g}_m$. This situation is so common that it may be illustrated by many different examples. Consider first a tank whose liquid level is to be controlled; its level responds with equal rapidity to inflow and outflow, one of which will be manipulated by the controller, while the other is the load. Most liquid-level and pressure loops have this same characteristic. Another example is the typical pH control loop; the stream to be neutralized and the manipulated reagent normally enter the vessel at the same point, thereby having the same dynamic effect on the controlled pH. Most blending systems also respond in this way.

Thermal systems are more complex, in that the load and manipulated variables are usually separated by a heat-transfer surface with different capacities on each side. In a typical heat exchanger, the heat capacity on the shell side will be greater than that on the tube side, although the deadtimes may differ very little. The same tends to be true of fired heaters and boilers, where the controlled temperature or pressure of the fluid leaving responds somewhat faster to a change in its flow (the heat load) than to a change in manipulated firing rate. Another source of load change to the fired heater is the heating value of the fuel itself, which has the same dynamic response as the fuel flow.

A distillation column is even more complex, having as many as four to six manipulated and controlled variables. There are two liquid levels to be controlled, column pressure, and one or more compositions, often measured as temperature. The composition variables are by far the slowest and most difficult to control. The response of a composition measurement is dominated by the same large time constant without regard to the choice of manipulated variable used to control it or to the source of the load disturbance. This seems to hold true for most lag-dominant processes and is always true where the process has an integrator.

In a multiloop setting such as distillation, there is usually a choice over the manipulated variable to be used to control the most critical variables, the quality of the products. All other manipulated variables then tend to act as load disturbances to the composition loops. It behooves the control engineer to base selection of the manipulated variables for composition control on a combination of high steady-state gain and fast response. Thus, if another variable has a significantly faster impact on the controlled variable, the engineer may have made the wrong choice and should restructure the system. In a properly structured system, then, $\mathbf{g}_q$ will not approach zero but is more likely to be equal to or slower than $\mathbf{g}_m$.

It will be convenient in most of the analysis that follows to start from the assumption that the load enters the process at the same point as the controller output, making $K_q = K_m = K_p$ and $\mathbf{g}_q = \mathbf{g}_m = \mathbf{g}_p$, where K_p and $\mathbf{g}_p$ are the steady-state and dynamic gains applied commonly to both inputs. Any actual difference from this assumption will be treated as a special case.

Disturbance shape

Because the magnitude of a dynamic-gain vector varies with the frequency of the signal passing through, the frequency content of the load disturbance has as much effect on the response of the loop as the vector itself. One of the possible disturbance shapes is a sine wave, either damped or undamped. This type of disturbance is most likely in a multiloop setting, where cycling is induced by another control loop. If the period of the disturbance is shorter than that of the loop it is disturbing, it may be attenuated somewhat by the dynamic gain vector $\mathbf{g}_q$ or $\mathbf{g}_p$ but not by the controller, which is powerless against faster disturbances. If the period of the disturbance is longer, the controller can be effective in attenuating it. The worst case is that of equal periods, since the closed-loop gain in that region tends to exceed unity, and the disturbance may actually be amplified. Identical interacting loops pose the greatest problem in this regard.

Any disturbance other than a pure sine wave will have a mixture of frequencies. A pulse has only high frequencies, a ramp has only low frequencies, and a step contains the entire spectrum from zero frequency, representing its final value, to the highest frequency, representing its leading edge. The step will then be recognized as the most difficult of the aperiodic disturbances. It therefore represents a good test of the controller response over its entire frequency range and at the same time is the easiest to administer. A step change in load may be simulated from a steady state simply by stepping the output of the controller while in manual, followed by transfer to automatic. The upset will drive the controlled variable away from set point, with the controller responding by returning its output back to its original steady state. This is, in fact, the preferred disturbance to use when tuning a controller, unless set-point response is to be optimized for some reason.

There may be specific instances where a pulse or ramp disturbance may be more representative of the demands on the process being controlled. In such cases, the controller should be tuned and tested using the relevant disturbance, since both the tuning and performance do depend on the shape of the load disturbance.

Load Response Curves

These curves are made of two sections: departure and recovery. The departure from set point is determined entirely by the size and shape of the load change and by the dynamics in its path. The departure section is therefore independent of the controller and will continue indefinitely in the absence of a controller. Recovery begins as soon as control action has an effect on the controlled variable, which is determined by the deadtime in the manipulated-variable path. The recovery section of the curve is therefore determined by the dynamics in the manipulated-variable path and the response of the controller to the departure section.

While the departure section is independent of the controller, there may be some argument over what constitutes the best possible recovery and hence the best possible controller. I arrived at the best possible recovery intuitively, then derived the controller output action that produced it, and finally tried to reproduce it using various controllers. The ability of some controllers to closely approach the best possible recovery (at least for first-order processes) but not surpass it confirmed the authenticity of the curves. Attempts to speed recovery by increasing control action always resulted in overshoot and oscillation and therefore extended the settling time and increased the area under the curve.

For step-load changes applied at the controller output, the response curves described below do not require the controller to saturate, i.e., to reach a limit. The reader may then argue that recovery could be hastened by saturating the controller output, in which case an unsaturated output does not produce the best *possible* recovery trajectory. There may, in fact, be a class of "hyperactive" controllers which do saturate on any deviation whatever. (On-off controllers behave this way but also leave the loop in a limit cycle, which disqualifies them from consideration as optimal load regulators.) However, no feedback controller commercially available at this writing can both saturate following a step load change applied at its output for a process having deadtime and return the loop to a true steady state afterwards. To avoid ruling out the possibility of future hyperactive controllers exceeding the performance limits described below, these limits will be called "best practical" where saturation may be desirable but not used and "best possible" where saturation is either fully developed or unnecessary. In the case of set-point steps, saturation is desirable and can be developed so that all limiting set-point–following trajectories will be classed as best possible.

The shape of the response curve is unaffected by the deadtime in the load path, so no error will be introduced by assuming equal deadtimes in the two paths. In the case where the process contains an integrator, it will be common to both paths. But *lags* appearing in the two paths may *not* be equal, and this will affect the shape of both the departure section and the best practical recovery curve. This imbalance is covered in detail under self-regulating processes below.

Pure deadtime

If the process being controlled contains no capacity whatever, a step change in load will produce a step change in the controlled variable one deadtime later, as described in Fig. 2.2. The deviation $c - r$ is normalized by including the process steady-state gain and the size of the upset in the scale so that the peak deviation falls at 1.0. The best that any controller can do for this process would be to step its output an amount equal to the load change as soon as possible, which is as soon as the deviation appears. Of course, the extent of the load change is unknown to the controller, which has only the deviation to act on with whatever gain it is given. If such an output trajectory is followed, however, the deviation will be restored to zero at the expiration of the next deadtime, which is as soon as possible. The maximum deviation for this best possible case is then

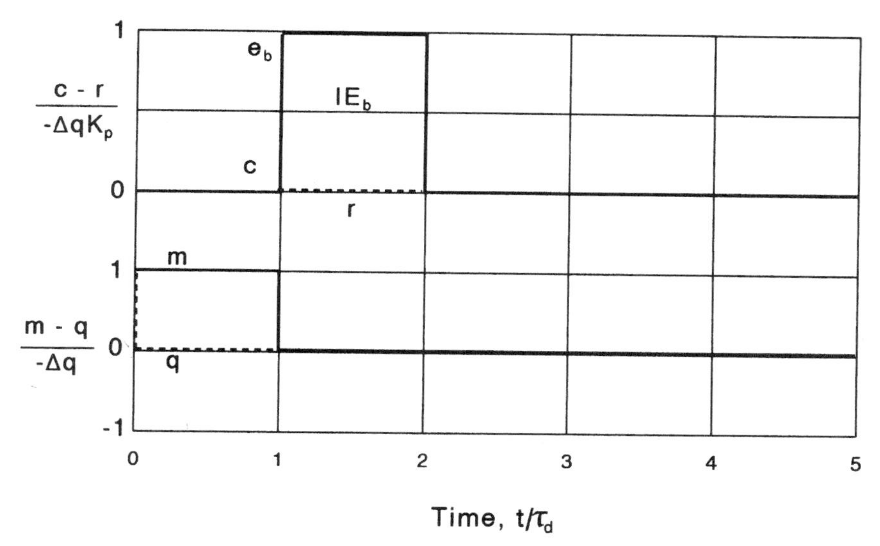

Figure 2.2 A step change in load to a deadtime process will produce a deviation that persists for at least a full deadtime.

$$e_b = -\Delta q K_p \tag{2.1}$$

and the integrated error is

$$\mathrm{IE}_b = e_b \tau_d \tag{2.2}$$

If the deadtimes in the two paths should differ rather than being common, the deviation would appear either earlier or later, but its shape and duration would not be affected, nor its area. Thus differing dynamics has no relevance to the pure-deadtime process. Saturation is not required here and would produce an overshoot if used.

Non-self-regulating processes

These processes contain an integrator, the controlled variable being the time integral of the difference between the load and the manipulated variable.[2] The most common example is the response of the liquid level in a vessel (indicating accumulated volume) to the difference between inflow and outflow. The name given to such processes indicates that without control, the measured variable such as liquid level will not find a steady state but will continue to ramp on any imbalance between flows until the vessel either empties or overflows. By contrast, a self-regulating process inherently seeks a steady state because its inflow and/or outflow is a function of the value of the controlled variable.

Most pressure loops are self-regulating, in that rising pressure tends to reduce inflow and increase outflow. Most thermal processes are also self-regulating, because a rise in process temperature tends to increase heat outflow and/or reduce heat inflow, stabilizing temperature. There are exceptions, however, as in exothermic reactors, where the heat produced by the reaction increases with temperature—these also will be covered.

Most liquid-level loops are non-self-regulating, however, because inflow is rarely affected by level and outflow may be controlled itself or determined by a pressure drop much higher than the head of liquid in the vessel. While composition loops such as pH are ordinarily self-regulating, recycling the discharged stream from the process to return unused ingredients to the feed removes that self-regulation. A non-self-regulating process is more difficult to control than its self-regulating counterpart and must be recognized as such by the control engineer. In fact, an integral-only controller cannot be used to control such processes because continuous oscillation will result.

Following a step change in load from an initial steady state, the non-self-regulating variable will ramp at a rate directly proportional to the size of the step and inversely proportional to its time constant τ_1. If the controller responds immediately to an appearance of a deviation, the effect of its action on the controlled variable will be delayed by the deadtime τ_d in the manipulated path, allowing the deviation to reach a value of

$$e_b = -\Delta q \, \tau_d / \tau_1 \tag{2.3}$$

as described in Fig. 2.3.

To restore the controlled variable to its original position, i.e., set point, the area between the manipulated and load variables developed before the controller acts must be canceled by an equal area with opposite sign afterward. This is the minimum requirement for a regulator. If the manipulated variable were only to duplicate the load trajectory—as it did with the deadtime process—the controlled variable would stop ramping at the end of the deadtime in the manipulated-variable path but remain at that value indefinitely. (It will be demonstrated later that this is the unfiltered output trajectory of an internal-model controller subject to a step change in load applied at the controller output for any process.)

The rate of return to set point is proportional to the difference between manipulated variable and load, i.e., the amount that the manipulated variable overshoots the load—no overshoot, no return. If the overshoot is 100 percent, the return to set point will proceed at the same rate as the departure, producing the isosceles triangle

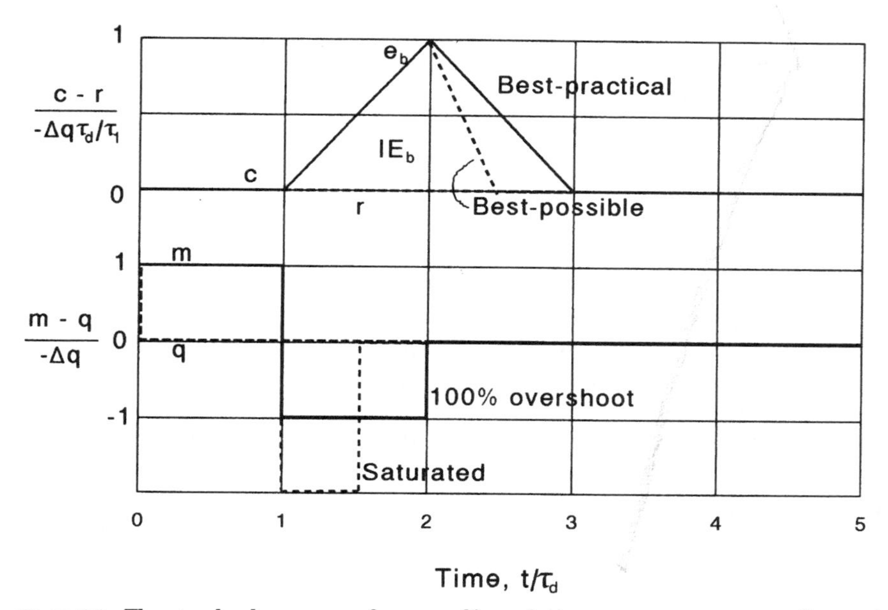

Figure 2.3 The step load response of a non-self-regulating process ramps away from set point for one deadtime; the best practical return ramp follows during the next deadtime.

appearing in Fig. 2.3. Saturating the controller output could produce a faster rate of return, but equal areas are still required on both sides of the load trajectory, requiring m to return to q before τ_d times out. While this action might be possible, it will be difficult to achieve, in that there is no change in the trajectory of the controlled variable to signal the time to make that correction, as is the case when the overshoot in m is 100 percent.

The integrated error under the best practical controlled-variable response curve may be calculated by multiplying the peak deviation by deadtime, as done in Eq. 2.2. Non-self-regulating processes always have load and manipulated variable entering the same integrator, so no consideration need be given to the possibility of unequal time constants.

Positive self-regulation

A single-capacity process having positive self-regulation (or simply self-regulation) responds to a step change with the familiar exponential response curve of a first-order lag:

$$\Delta c = -\Delta q K_p(1 - e^{-t/\tau_1}) \tag{2.4}$$

where t is the time from the beginning of the response, and τ_1 is the time constant of the lag. For the departure section of the response curve, the gain and lag would be those in the load path.

As before, if the controller acts as soon as the deviation appears, its impact on the controlled variable will take place following the deadtime in the manipulated path, at which point the maximum deviation will have been reached:

$$e_b = -\Delta q K_q (1 - e^{-\tau_d/\tau_q}) \tag{2.5}$$

The best practical return to set point follows during the next deadtime, as shown in Fig. 2.4, along a trajectory that is the complement of the exponential departure trajectory. Both departure and recovery sections of the response curve are functions of the time constant τ_q in the load path and the deadtime τ_d in the manipulated path. As a result of the complementarity of the two curves, IE_b is once again the product of e_b and τ_d, as in Eq. 2.2.

To achieve this recovery, the controller output must overshoot the load. If the lags in the two paths are equal, the overshoot is less than required for the non-self-regulating process:

$$\frac{\Delta m}{\Delta q} = 1 + e^{-\tau_d/\tau_1} \tag{2.6}$$

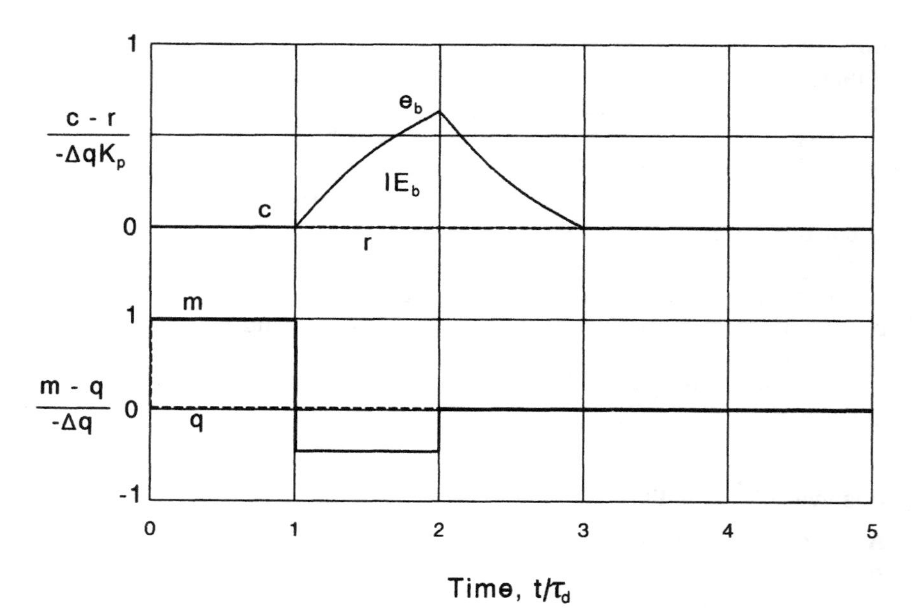

Figure 2.4 The best practical load response for processes with self-regulation requires some overshoot of the manipulated variable.

where τ_1 is common to both paths. Although the no-overshoot control action taken in Fig. 2.2 will eventually return the deviation to zero, recovery will take place along an exponential lag equal to the process time constant. This obviously extends the settling time of the controlled variable as well as increasing its integrated error as follows:

$$\frac{IE_m}{IE_b} = \frac{1}{1 - e^{-\tau_d/\tau_1}} \tag{2.7}$$

where subscript m designates the curve produced by matching the manipulated variable to the load rather than overshooting it. The difference between these two recovery curves increases with the ratio of τ_1/τ_d.

If the lag τ_q in the load path is shorter than τ_m in the manipulated path, the overshoot must be much greater to approach the best practical recovery curve in proportion to τ_m/τ_q. If this ratio is quite large, the overshoot could easily cause saturation on a small load change, which will interfere with recovery.

Negative self-regulation

A process having negative self-regulation is called *steady-state unstable*. A visual example is the inverted pendulum, which can be balanced but will accelerate away from the vertical upon the least disturbance. As mentioned earlier, exothermic chemical reactors can be steady-state unstable if on a temperature rise, heat produced by the reaction increases more than heat removed by the cooling system. Most exothermic batch reactors are steady-state unstable because of a large ratio of volume to heat-transfer surface.

This process may be said to have negative resistance, consistent with its negative self-regulation; its time constant is also negative. Still, such a process is controllable if its deadtime is significantly shorter than its time constant.[3] Dynamics in the load and manipulated paths tend to be similar in these processes.

The departure section of the response curve shown in Fig. 2.5 drives away from set point on an exponential trajectory having a positive sign in the exponent where Eq. 2.4 has a negative sign. However, its negative time constant produces this behavior so that Eqs. 2.4 through 2.6 apply equally to positive and negative self-regulation. The controller output must overshoot the load for this process even more than 100 percent or the controlled variable will run away. For this reason, cooling systems for exothermic reactors should be operated with plenty of margin—having adequate cooling for steady-state operation will not be enough to prevent a runaway following an upset.

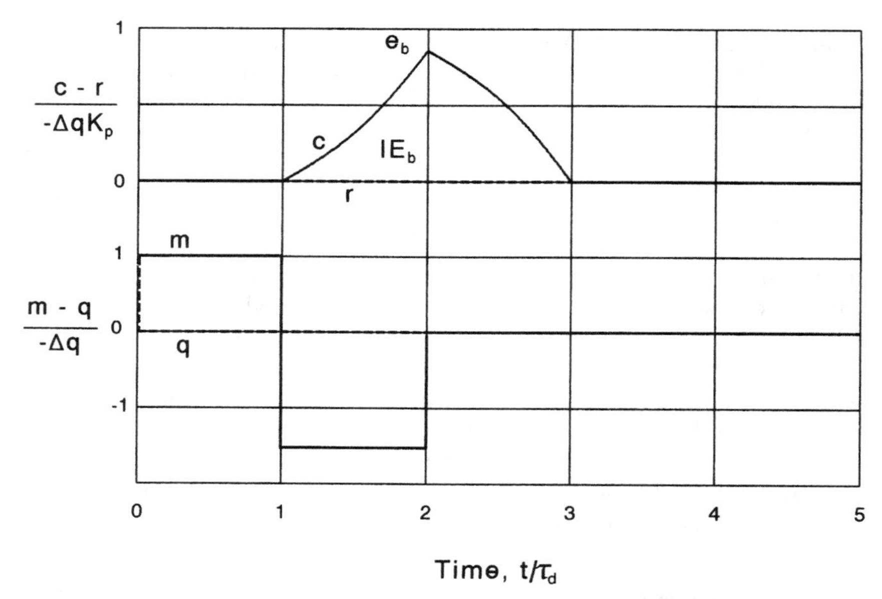

Figure 2.5 The unstable steady-state process accelerates away from equilibrium.

Effects of a secondary lag

Consider controlling a non-self-regulating process having a secondary lag in the measurement of the controlled variable, which would be common to both load and manipulated inputs. This would slow the response of c to both q and m, stretching the response curve and flattening its peak, as shown in Fig. 2.6. Surprisingly, the area under the response curve is unaffected by the secondary lag, remaining as it was for the first-order non-self-regulating process:

$$\mathrm{IE}_b = -\Delta q\, \tau_d^2/\tau_1 \tag{2.8}$$

and the peak height may be calculated from it:

$$e_b = \mathrm{IE}_b/(\tau_d + \tau_2) \tag{2.9}$$

A similar procedure may be followed for self-regulating processes, calculating IE_b as if the process were first-order:

$$\mathrm{IE}_b = -K_p \Delta q\, \tau_d(1 - e^{-\tau_d/\tau_1}) \tag{2.10}$$

and then estimating peak height using Eq. 2.9.

The best practical response curve for the second-order non-self-regulating process takes on the appearance of a symmetrical gaussian distribution curve. For the self-regulating process, the curve is less

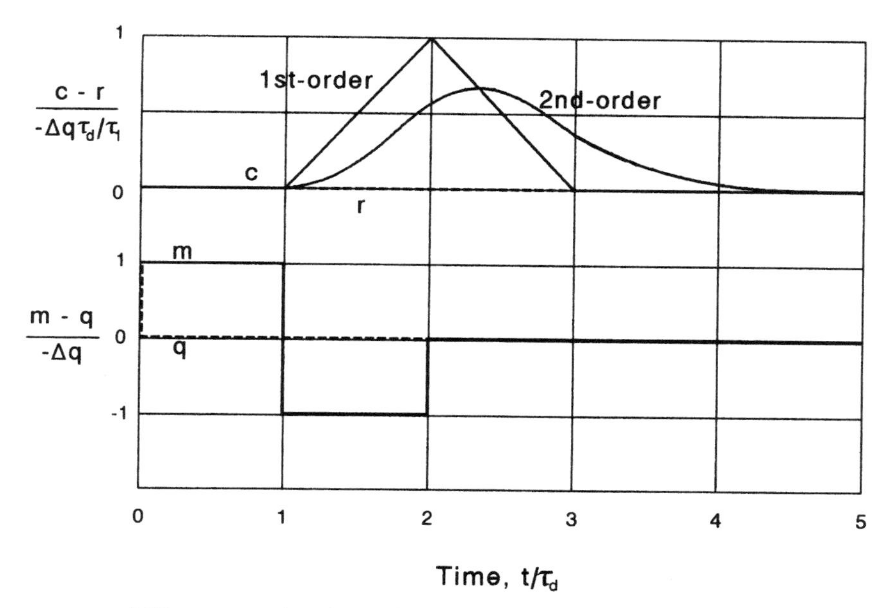

Figure 2.6 Adding a secondary lag reduces and delays the peak but does not affect the area of the best practical response curve.

symmetrical when deadtime dominates and more symmetrical as the lag dominates.

The reader should appreciate that while the secondary lag has no effect on IE_b, it makes that target more difficult to approach by a given controller. The secondary lag would have to be canceled by an identical lead in the controller to produce the area achieved when controlling a first-order process with deadtime. Since a lead cannot be produced without an accompanying lag, the actual IE achieved by a controller must increase with each additional process lag.

In the absence of deadtime, both IE_b and e_b are theoretically zero. Thus a multiple-lag process containing no real deadtime is theoretically capable of perfect control, given the caveat in the preceding paragraph. However, in a practical sense, any process containing three or more lags cannot be controlled perfectly, nor even identified perfectly.

Multiple noninteracting lags

Noninteracting lags are physically isolated from each other so that their overall response is simply that of the serial combination of the individual lags. A common example is the set of trays in a distillation column, wherein liquid cascades from the top to the bottom tray in a series of free falls, as shown in Fig. 2.7. Should the flow from a lower

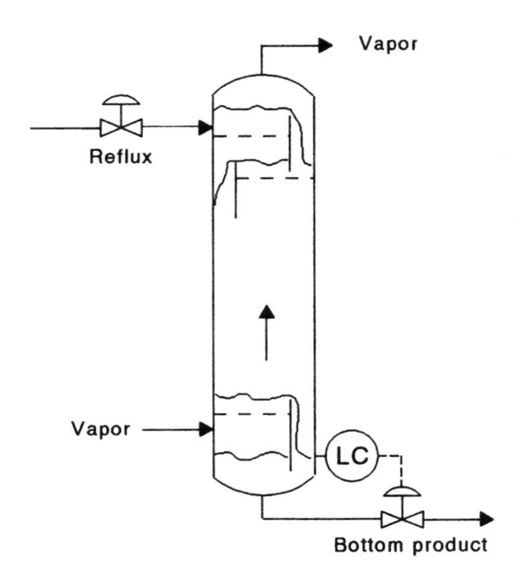

Figure 2.7 A distillation column has a series of noninteracting hydraulic lags between reflux and base level and a series of interacting lags in the response of compositions to reflux and vapor rates.

tray be restricted, this has no effect on the flows leaving any of the upper trays because of the decoupling effect of the free fall between them.

The practical limit for the step-load response of a process consisting of a series of equal noninteracting lags τ can be simulated by setting $m = q$ at the earliest indication of a response in the controlled variable. This assumes that the best trajectory for m has no overshoot, which becomes more accurate as the number n of lags increases. Figure 2.8 shows these responses for several values of n; note that its time scale is normalized to $n\tau$, which represents the total time response of the system. As n increases, the peak height and integrated error of the curves also increase, with 100 percent response being approached as n reaches 100. As n approaches infinity, response would be equivalent to pure deadtime. Figure 2.9 summarizes this relationship.

Multiple interacting lags

In a series of multiple interacting lags, information flows in both directions. The same distillation column used to illustrate noninteracting lags between flow and flow has interacting lags between flow and composition. An analyzer measuring the composition of the overhead vapor would respond to a change in reflux flow entering the top tray as a first-order lag. However, such a flow change also affects the composition of the liquid, leaving the top tray as well as its flow, which then change the composition of the vapor passing from the second to the top tray. This produces a secondary effect on the composi-

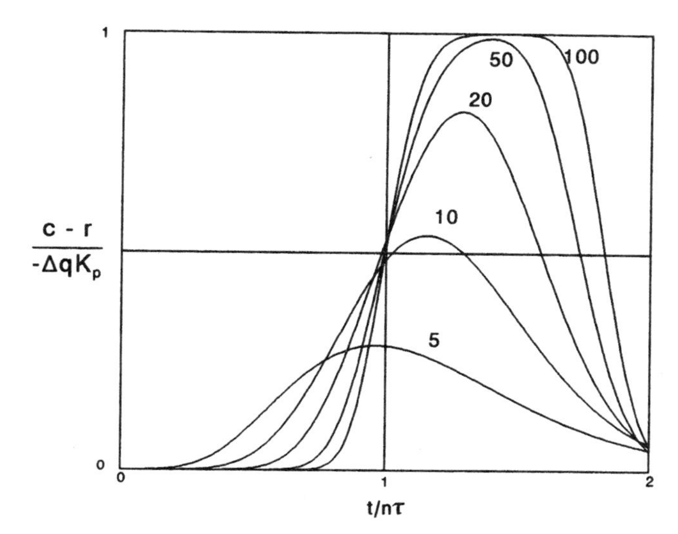

Figure 2.8 These step-load responses for multiple equal noninteracting lags were obtained by setting m equal to q when the response reached 2 percent.

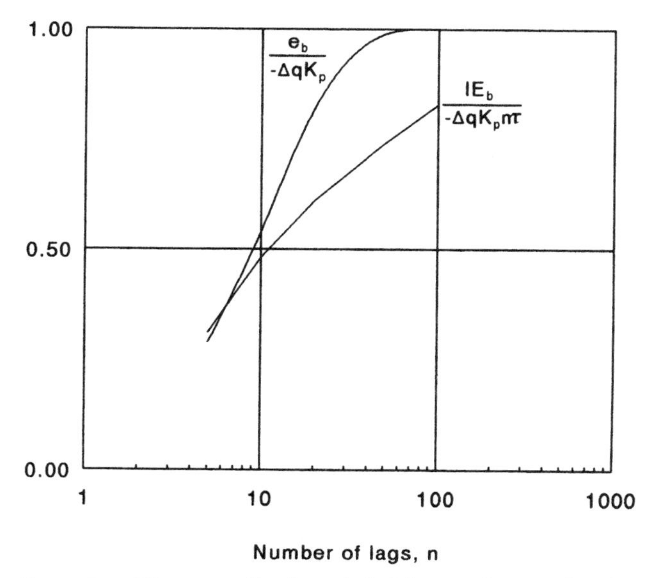

Figure 2.9 The normalized peak heights and areas for the curves appearing in Fig. 2.8 vary with the logarithm of n.

tion of the overhead vapor. As the liquid flow and composition on each lower tray change, the vapor composition leaving that tray is thereby changed; this subsequently affects liquid and vapor compositions on all upper trays, as well as that of the overhead vapor. As a result, all trays gradually approach equilibrium together.

This interaction produces a very different step response than is observed with noninteracting lags. Instead of the response approaching that of pure deadtime as n approaches infinity, one dominant lag remains, accompanied by a much shorter deadtime and secondary lag. Open-loop step-response curves for processes having 100 interacting and noninteracting lags, respectively, are compared in Fig. 2.10. The time scales for the two curves differ as well as the shapes. The time $\Sigma\tau$ at midscale is $n\tau$ for the noninteracting lags and $\Sigma n\tau$ for the interacting lags,[4] where Σn is the sum of the numbers from 1 to n; it may be calculated alternatively as $(n^2 + n)/2$. For the 100-lag process, Σn is 5050. Observe that the curve for the noninteracting lags crosses 50 percent response at time $\Sigma\tau$ and that of the interacting lags crosses 63.2 percent response $(1 - e^{-1})$, as does that of a first-order lag.

The dynamic response of multiple interacting lags can be represented quite satisfactorily by a dominant primary lag along with a smaller deadtime and secondary lag. Table 2.1 gives their values relative to $\Sigma\tau$ as a function of n as identified by pulse testing. This model allows estimates of IE_b and e_b to be made using Eqs. 2.10 and 2.9; $\Sigma\tau$

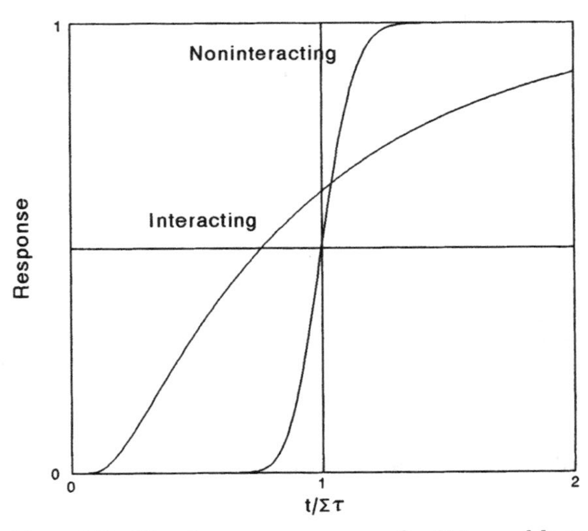

Figure 2.10 The step-response curves for 100 equal lags reveal a much lower effective deadtime when the lags interact.

TABLE 2.1 Modeling of *n* Equal Interacting Lags

n	$\tau_d/\Sigma\tau$	$\tau_1/\Sigma\tau$	$\tau_2/\Sigma\tau$
1	0	1.000	0
2	0	0.873	0.127
3	0.043	0.828	0.129
5	0.075	0.796	0.129
10	0.095	0.790	0.115
20	0.105	0.790	0.105
50	0.105	0.790	0.105
∞	0.105	0.790	0.105

can be estimated from an open-loop step-response curve as the time required to reach 63.2 percent response.

Distributed lags

A very important class of processes has the dynamic character of a *distributed lag*. A transmission line, for example, whether electric or pneumatic, has both resistance and capacitance distributed uniformly across its entire length rather than lumped at discrete locations. Some distillation columns have packing rather than trays, which distributes their capacity. And all heat exchangers have both heat-transfer surface and heat capacity distributed over their entire length.

These distributed processes may be modeled as an infinite series of infinitesimally small interacting lags and as such will have a step-response curve indistinguishable from that of the 100 interacting lags appearing in Fig. 2.10. This behavior has been verified by tests conducted on heat exchangers, boilers, packed columns, and pneumatic transmission lines.

The response of the composition of a blend of two or more ingredients within or leaving a vessel to a step change in the proportions of those ingredients depends on whether the vessel is backmixed. In a plug-flow reactor, for example, backmixing is avoided, since conversion of feed to product during a given residence time is lowered by backmixing the product with the feed.[5] Its step-response curve has the shape of that of a large number of noninteracting lags—the less the mixing, the closer it approaches pure deadtime. By contrast, the backmixed reactor responds as a series of interacting lags. The two curves appearing in Fig. 2.10 are quite representative of the two types of reactors, with time at midscale being the residence time of the reactor:

$$\Sigma\tau = V/F \tag{2.11}$$

where V is the volume of the mixture, and F is its flow rate.

The apparent deadtime depends on the thoroughness of backmixing, expressed as the ratio of the circulation rate F_a of the mixer to the feed rate:[6]

$$\tau_d/\Sigma\tau = \frac{1}{2(1 + F_a/F)} \tag{2.12}$$

There also would be a secondary lag comparable with the deadtime, as shown for the high-order processes in Table 2.1; the dominant time constant would be $\Sigma\tau$ less τ_d and τ_2, as expected.

A true distributed lag such as a transmission line consisting of a distribution of resistance and capacitance uniformly along its length has the following transfer function:

$$G(s) = \frac{2}{e^{\tau s} + e^{-\tau s}} = \frac{1}{\cosh\sqrt{\tau s}} \tag{2.13}$$

where s is the LaPlace operator.

The hyperbolic cosine may be expanded into an infinite-product series containing the roots of the function:

$$G(s) = \frac{1}{[1 + (2/\pi)^2\tau s][1 + (2/3\pi)^2\tau s][1 + (2/5\pi)^2\tau s]\cdots} \tag{2.14}$$

The first term in the series is a dominant lag whose time constant is $(2/\pi)^2\tau$ or 0.405τ; the second has a time constant of $(2/3\pi)^2\tau$ or 0.0450τ. The sum of all the time constants is 0.500τ. All the remaining time constants are progressively smaller and may be lumped as deadtime having the value of the sum less the first two, or 0.050τ. These values compare favorably with the model given in Table 2.1 for $n > 20$, in that $\tau_d/\Sigma\tau = 0.100$, $\tau_1/\Sigma\tau = 0.810$, and $\tau_2/\Sigma\tau = 0.090$.

Set-Point Response Curves

Set-point changes are easier for a controller to follow than are load changes for two reasons:

1. The controller has access to precise set-point information, whereas the load is unknown.
2. The set point enters the controller either directly or through a known filter, whereas the load is always filtered by process dynamic elements which are generally unknown and may be variable.

As a result, the controller may be made to respond to set-point changes by saturating if desirable, and set-point response can be shaped to any extent less severe than the limits given below.

First-order processes

For a pure deadtime process, the absolute limit for set-point response is for the controlled variable to reproduce the set-point trajectory precisely, one deadtime later. This does not require saturation of the controller output, simply needing it to step immediately by $\Delta m = \Delta r / K_p$. The best possible integrated error is

$$\mathrm{IE}_b = \Delta r \tau_d \tag{2.15}$$

Where deadtime accompanies a lag or integrator, the preceding contribution must be added to values of IE_b calculated for first-order processes without deadtime, given below.

With an integrator in the process, the controlled variable may be made to ramp toward a new set point at a rate directly proportional to the difference Δm_1 between the saturated controller output and the load and inversely proportional to the integrator's time constant. Saturation is then maintained for the time interval Δt_1 required to reach the new set point:

$$\Delta t_1 = |\Delta r\ \tau_1 / \Delta m_1| \tag{2.16}$$

The deviation integrated during that time is half the base of the triangle shown in Fig. 2.11 multiplied by its height:

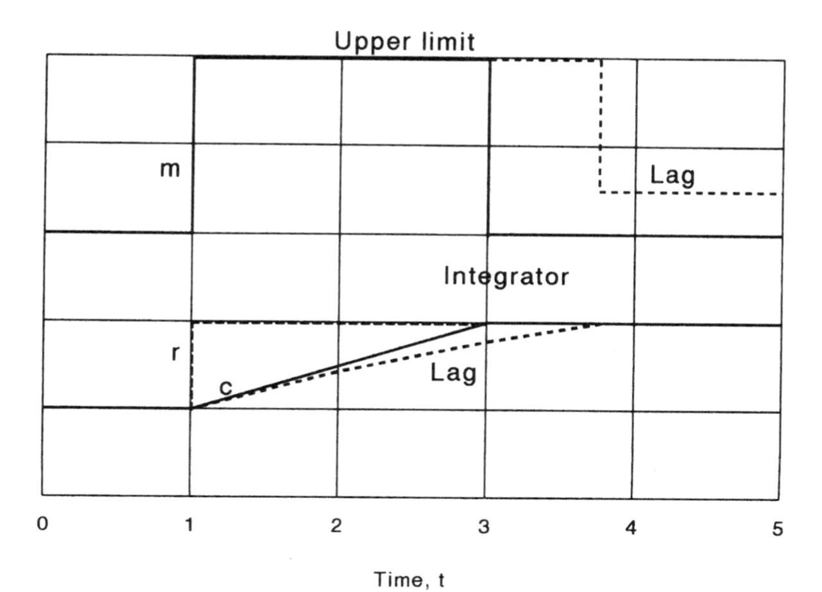

Figure 2.11 The best possible set-point response for a first-order process requires saturation for the time required to reach the new set point.

$$\mathrm{IE}_b = \Delta r \Delta t_1/2 \tag{2.17}$$

In the presence of deadtime, the contribution from Eq. 2.15 must be added, and m must be returned to equal the load, one deadtime before the new set point is reached.

For a first-order self-regulating process, the controlled variable follows an exponential trajectory terminated at the new set point, also shown in Fig. 2.11. The time required to reach the new set point (also the time during which saturation is maintained) is

$$\Delta t_1 = -\tau_1 \ln(1 - \Delta r/\Delta m_1 K_p) \tag{2.18}$$

where the leading negative sign applies because the logarithm is always negative.

Integrating the deviation between the set point and the exponential response curve gives

$$\mathrm{IE}_b = \Delta r[\tau_1 + (1 - K_p \Delta m_1/\Delta r)\Delta t_1] \tag{2.19}$$

The solution to Eq. 2.19 approaches that of Eq. 2.17 as $\Delta r / K_p$ approaches zero. Again, add the contribution of any deadtime.

Observe that in Fig. 2.11, m terminates in a new position for the self-regulating process rather than returning to its previous position, as required by the integrating process. The difference in m between the two steady states is $\Delta r / K_p$. The steady-state value of m in the self-regulating process includes components of both set point and load, because self-regulation depends on process inflow and/or outflow changing with the value of the controlled variable.

Effects of secondary lags

The presence of a second-order lag complicates the issue considerably. Best possible IE (minimum time) now requires the controller to saturate twice—first to accelerate the controlled variable toward the new set point and then at the opposite limit to decelerate and thereby avoid overshoot. If the second saturation is not used, m must leave the first limit earlier (lacking the deceleration above), extending the time to reach the new set point. Although saturation at both limits minimizes the time required to reach the new set point—and hence IE—it uses more resources such as energy. Conversely, the minimum-resource trajectory does not minimize time.

For the minimum-resource trajectory, shown in Fig. 2.12, approach to the new set point is exponential, with IE increasing by the following amount, using the same formula as for deadtime:

$$\Delta \mathrm{IE} = \Delta r \tau_2 \tag{2.20}$$

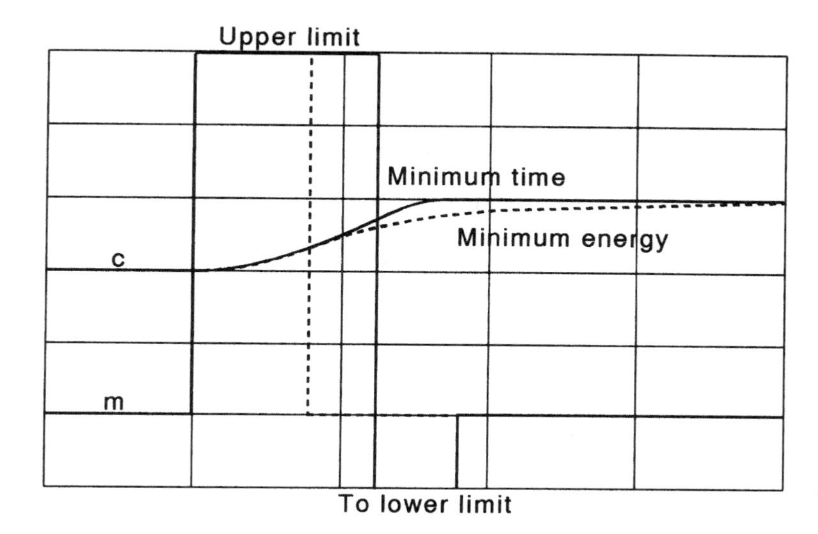

Figure 2.12 For a second-order process, minimum energy use requires saturation once, while minimum time requires saturation at both limits.

This contribution must be added to that of the primary time constant and any deadtime. The time during which m is held at saturation is the same as for a first-order process.

For the minimum-time trajectory, the second saturation must be held for a duration of

$$\Delta t_2 = -\tau_2 \ln |\Delta m_2| \tag{2.21}$$

where Δm_2 is the difference between the second limit and the final steady-state value of m. Now, the duration of the first saturation must be extended by $\Delta t_2(-\Delta m_2/\Delta m_1)$ to provide the additional energy or other resource consumed by the second saturation. The total time required to reach the set point is then the sum of the three time intervals:

$$\Delta t = \Delta t_1 + \Delta t_2(1 - \Delta m_2/\Delta m_1) \tag{2.22}$$

The result of the second saturation is to cut off the exponential approach to set point, as shown in Fig. 2.12, terminating the curve at the time estimated in Eq. 2.22. This surgery reduces the IE contribution of the secondary lag to approximately

$$\Delta IE_b = \tau_2(\Delta r - \Delta m_1 \tau_2/2\tau_1) \tag{2.23}$$

If the process load is zero (e.g., a batch process), then Δm_2 also may be zero, in which case Eq. 2.21 has no solution. The minimum-time and minimum-resource programs are identical under this condition.

Performance and Robustness Measures

In selecting and tuning a controller, one should be guided by measures of both performance and robustness. These measures will vary with the type of controller selected, the nature of the process being controlled, the source of the disturbance, the value of any sampling element or filter in the loop, and, of course, the tuning of the controller. As best practical performance is approached, it is axiomatic that robustness approaches zero. While robustness can be improved by filtering and detuning, performance is thereby compromised. Only through the use of absolute measures of performance and robustness can one make informed decisions about controller selection and tuning.

Having established limits for best possible (or best practical) performance, it is only necessary to divide by the actual performance measured for the same criterion and in the same units to produce a dimensionless performance index. While any of the criteria listed in Chap. 1 could be used to measure controller performance, the integrated-error criterion seems to be the most useful, as well as the most economically significant. On this basis, controller performance is simply IE_b/IE. Other criteria could be used as well, such as IAE or peak deviation, depending on the control objective. However, the criterion must be identified along with the performance index. Unless otherwise noted, the IE performance index is used throughout the remainder of this book.

Control-loop robustness

To this point, performance has been the principal concern. However, robustness is equally important in a plant situation, because a loop that lacks it cannot be left closed and therefore cannot benefit from the performance expected of the controller. *Robustness* is a measure of the smallest change in any process parameter (from conditions at which the controller was tuned) that will cause an undamped oscillation to be sustained in the loop.

At least two of the following process parameters will affect the stability of a given control loop: K_p, τ_d, τ_1, or τ_2. To identify the variation in two of the parameters individually or in combination which will lead to instability, the two-dimensional robustness plot appearing in Fig.

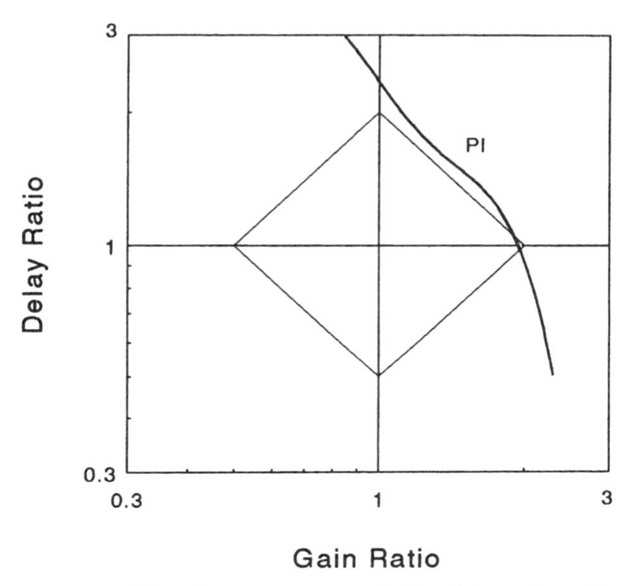

Figure 2.13 The robustness measure is the distance from the center of the plot to the closest point on the stability-limit curve, here 0.97 for PI control of a deadtime process.

2.13 has been devised. In this example, the two parameters investigated are τ_d (delay ratio) and K_p (gain ratio). The center of the coordinate system represents the conditions prevailing when the controller was tuned. Moving to the right represents an increase in K_p, while moving up represents an increase in τ_d. Logarithmic coordinates are used to allow treatment of gain and deadtime changes equally in both directions. The diamond-shaped window is the locus of all points having a combined factor of 2 change in both deadtime and gain in any direction. It serves simply as a frame of reference for a robust system whose stability limit lies outside the window.

To demonstrate the use of the robustness plot, the stability limit envelope for a loop having a PI controller and a pure deadtime process has been drawn on the coordinate system. This envelope is the locus of all points where a uniform oscillation was produced by changes in K_p and/or τ_d. The controller was first tuned for minimum IAE in response to a step load change. Then K_p was increased until a uniform oscillation developed—this occurred at a 97 percent increase (gain ratio 1.97). With the gain returned to its original value, τ_d was increased until uniform oscillation again developed—this occurred on a 130 percent increase (delay ratio 2.3). Then for a fixed increase in τ_d, K_p also was increased until a uniform oscillation developed, with each combination plotted as a point on the coordinate system. The stability-limit envelope is the curve connecting all those points.

A single measure of robustness is the smallest combination of changes in the two process parameters that can precipitate instability. For the example shown, this point on the envelope happens to fall at a gain ratio of 1.97, just cutting across the right corner of the window. The distance from the center to this point is 0.97, the robustness index for the example loop.

Should the process being controlled have three or four variable parameters, the robustness plot should have as many dimensions. In practice, however, it will be easier to prepare two-dimensional plots for combinations of these parameters taken two at a time.

However, in every case examined by me, the closest point on the stability-limit envelope fell on one axis or the other, as it did in Fig. 2.13. In this light, it may not then be necessary to vary the process parameters in combination but simply to change them individually until instability is reached and report the smallest of the changes required. This procedure reduces the number of tests to a minimum and avoids the need for plotting at all. While the robustness plot is useful as a tutorial, it may not be required in analysis, unless later experience with more complex loops indicates otherwise.

Notation

c	Controlled variable
d	Differential operator
e	Deviation, exponential operator 2.718
e_b	Lowest possible peak deviation
F	Feed rate
F_a	Recirculation rate
$\mathbf{g}$	Dynamic gain vector
IE	Integrated error
IE_b	Lowest practical integrated error
IE_m	Integrated error produced by setting $m = q$
IAE	Integrated absolute error
K	Steady-state gain
m	Manipulated variable
Δm_1	Difference between upper limit and load

Δm_2	Difference between lower limit and load
n	Number of lags
Σn	Sum of the numbers from 1 to n
q	Load variable
r	Set point
s	LaPlace operator d/dt
t	Time
Δt_1	Duration of first saturation
Δt_2	Duration of second saturation
V	Volume
Δ	Change
Σ	Summation
τ	Time constant
τ_d	Deadtime
τ_1	Primary time constant
τ_2	Secondary time constant
$\Sigma\tau$	Total response time

Subscripts

m	Manipulated variable
p	Process (common to load and manipulated variable)
q	Load variable

References

1. Shinskey, F. G., *Process Control Systems,* 3d Ed., McGraw-Hill, New York, 1988, pp. 22, 34, 127–129.
2. Ibid., pp. 23, 24.
3. Ibid., pp. 181–186.
4. Ibid., p. 47.
5. Ibid., pp. 375, 376.
6. Ibid., pp. 99–102.

Part

2

Linear Controllers

Chapter

3

PID Controllers

The bulk of the control work in the process industries is done by members of the proportional-integral-derivative (PID) control family, and this will continue to be true into the future owing to their combination of simplicity, familiarity, and robustness. And the most prominent member of the family (in numbers) is the proportional-integral (PI) controller, which is the simplest configuration combining stability (proportional) with elimination of steady-state offset (integral). Over the years, this family has grown from simple mechanical and pneumatic devices to complex digital algorithms replete with all the "bells and whistles" imaginable. This chapter will concentrate on the functionality of these controllers, carefully distinguishing the many different offerings on the basis of operational features, mode interaction, and suitability for given applications.

This application suitability has a special importance with respect to the use of integrating action. While most applications benefit by integrating action, with some it is absolutely detrimental to achieving the control objective. These cases will be identified as they arise, and the appropriate solution will be provided. Among other things, the use of derivative action will be encouraged, because of the improvement in performance it provides on lag-dominant processes. Derivative action is not used nearly enough in the field, perhaps because it is not well understood, especially with regard to tuning.

Proportional Control

The simplest—but not necessarily the best understood—of the PID controllers is the proportional controller. The position m of its output is related proportionally to the deviation between the set point r and the controlled variable c:

$$m = \pm\frac{100}{P}(r - c) + b \tag{3.1}$$

where the ± sign indicates the choice of controller action required to produce a negative-feedback loop. The action most commonly used calls for an increase in the controller output upon a decrease in the controlled variable (increase-decrease action), which would correspond to a + sign in Eq. 3.1.

The steady-state gain of the controller identified as K_c in Fig. 2.1 is its proportional gain, expressed in Eq. 3.1 as the inverse of its proportional band P in percent. Controller manufacturers are split about evenly between the use of proportional gain and percentage proportional band. When gain is used, it is ordinarily a pure number, in that controller input and output are both expressed in percentage of full scale. However, early work by Ziegler and Nichols[1] report gain as "sensitivity" in units of "psi/in." One had to know that the output span of the controller was 12 psi (lb/in^2) and that the input span was 4 in of pen travel to convert their results to dimensionless gain.

Proportional band P, also known as *throttling range,* is defined as the change in the controlled variable that will drive the controller output full scale. For most controllers using proportional band, it is expressed in terms of percent, as is the case in Eq. 3.1. However, some special-purpose controllers—especially those used to control temperature—may have their proportional band expressed in such units of measure as degrees Celsius. The proportional gain of such a controller would then be calculated as input span/P, just as in Eq. 3.1 input span is 100 percent. For ease of presentation, 100/P will be used to identify the nominal proportional gain throughout this book, which also will be the true proportional gain K_c, *except* in the case of the interacting PID controller, where their differences will be identified.

Because this class of controllers is universally referred to as *PID,* meaning "proportional, integral, and derivative," the controller settings used in this book will be expressed in terms of proportional band, integral time, and derivative time unless specifically noted otherwise.

Bias, offset, manual reset

Equation 3.1 indicates that deviation between r and c will be zero *only* when output m happens to equal output bias b. As a result, any changes in either load or set point requiring m to settle at a different steady-state value will produce a corresponding deviation:

$$e = \pm\frac{P}{100}(m - b) \tag{3.2}$$

where $e = r - c$. The deviation produced is known as *proportional offset*. If the proportional band is very narrow or the difference between m and b is small, then offset will be small. This allows proportional controllers to be used for such applications as liquid-level control, where P can be as low as 10 percent, and no particular economic penalty applies to small deviations in level.

Before integral action was added to eliminate offset (*ca.* 1929), it was common practice for operators to apply their own correction by repositioning or *resetting* the set point. While this action did not eliminate offset, it did move the controlled variable by the same amount the set point was moved. Alternatively, the controlled variable could be moved by manually adjusting the bias. This could eliminate offset, but the bias adjustment required was not as predictable as the amount of set-point adjustment. Both practices were known as *manual reset*. Not surprisingly, *automatic reset* (integral action) involved automatically repositioning the bias of the controller.

Now, integral action is so common that some product lines have no strictly proportional controller. While one can be approximated by turning off or dialing out integral action, this may not be sufficient to solve the problem at hand. If no fixed bias is available, the controller in question is not a true proportional controller. Simply turning off integral action usually leaves its constant of integration as the output bias for the controller, and this may float. If the integral action cannot be stopped completely, the constant of integration will drift with time and deviation. For most controllers, however, it also will change whenever the controller is transferred from manual to automatic operation. To avoid bumping the output on such a transfer, most controllers back-calculate the bias by solving Eq. 3.1 for b in terms of the present values of m (as introduced manually) and e at the time of transfer. Even if b were to be carefully set by positioning the output of the controller manually relative to observed deviation and proportional band, any later operator intervention would change it. Consequently, it is not advisable to try to make a proportional controller out of a PI controller unless there is some sure means of inserting a fixed bias.

Zero-load processes

A very important application requiring a fixed bias in either a proportional or proportional-derivative controller is the *zero-load process*. This is usually identified as a batch operation, where there is no flow through the process being controlled. For example, a tankful of material may need adjustment in composition or temperature before the next operation is to be conducted on it. There being no flow through the vessel, if any more than the precise amount of reagent or heat than required is added to it to reach set point, a permanent overshoot will result. An example is the familiar laboratory titration, where if too much reagent is added, the endpoint will be passed, and there is no remedy. Integral action applied to the controller in this loop will always result in overshoot.

Here the bias can be safely set at zero, because the process has that as its steady-state load. When the set point is reached, the correct value of the manipulated variable is zero. While in some cases the opposite reagent also may be available or cooling can be manipulated in split range with heating, this only means that any overshoot is not permanent. It does not change the requirement for the fixed bias—b should still be set so that the manipulated variable is shut off when the deviation is zero.

Some processes *approach* zero load without quite reaching that point. An example is the heating of a large mass of cold-rolled steel in an annealing furnace. The fuel flow must be maximized to raise the furnace temperature as quickly as possible to set point, but keeping it at set point requires only enough heat to offset losses to the surroundings. While this load is not quite zero, it is small and predictable, so integral action is not required.

Averaging-level control

A very common application mentioned briefly in Chap. 1 is the smoothing of the flow of a feedstream to a process by using the capacity of its feed tank. There are two objectives to be met: to minimize changes in the flow of the manipulated stream and to keep the tank from overflowing and emptying. A flow controller is usually applied to maintain a constant and accurate feed rate to the downstream process. This guarantees that the tank level is non-self-regulating because inflow to the tank (the load) is not affected by level either. The manipulated flow and the load will not match without proportional action by the level controller, and yet they must be made to match if a steady state is ever to be reached.

Several controllers have been tried unsuccessfully to control this process. The most common is a standard PI controller with a midscale set point. While this loop can be made quite stable, stability depends heavily on a high proportional gain. Unfortunately, a high proportional gain is at cross-purposes with minimizing changes to the manipulated flow. As a result, the proportional gain is usually reduced below 1.0 (proportional band > 100). This combination of low proportional gain plus integration acting on an integrating process tends to produce lightly damped oscillations having an excessively long period.

To illustrate the stability problem, a closed loop will be formed with an integrating process and a PI controller and reduced mathematically to the quadratic formula. Consider the liquid-level process as an integrator, represented mathematically as

$$\frac{dc}{dt} = \frac{q - m}{\tau_1} \tag{3.3}$$

where c is the liquid level, q is the independent inflow, m is the manipulated outflow, and the time constant of the process τ_1 is the capacity of the vessel over the span of the level measurement divided by the span of the flowmeter.

The PI controller may be described in its derivative form as

$$\frac{dm}{dt} = \frac{100}{P}\left[\frac{dc}{dt} + \frac{(c - r)}{I}\right] \tag{3.4}$$

where r is the set point, and I is the integral time constant. Taking a second derivative of c from Eq. 3.3 yields

$$\frac{d^2c}{dt^2} = -\frac{dm}{\tau_1 dt} \tag{3.5}$$

Substituting for dm/dt from Eq. 3.5 into Eq. 3.4 produces the following quadratic:

$$\left(\frac{\tau_1 IP}{100}\right)^2 \frac{d^2c}{dt} + I\frac{dc}{dt} + c = r \tag{3.6}$$

The coefficient of the second derivative is $\tau_n/2\pi$; solving for the undamped natural period τ_n gives

$$\tau_n = 2\pi\sqrt{\tau_1 IP/100} \tag{3.7}$$

The coefficient of the first derivative is $\zeta\tau_n/\pi$; solving for the damping factor ζ gives

$$\zeta = 0.5\sqrt{100I/\tau_1 P} \tag{3.8}$$

From Eq 3.8 we see that the damping factor *decreases* as P increases—the loop becomes *less* stable with increasing proportional band. This is counterintuitive and is the source of much confusion as well as instability. The period also increases with P, producing the slow, lightly damped cycles shown in Fig. 3.1. In the simulation used to produce the cycles, P was set at 200 percent and I at $\tau_1/5$, yielding values of $\tau_n = 4.0\tau_1$ and $\zeta = 0.16$.

Various nonlinear functions also have been applied to improve PI averaging-level control, from the error-squared controller (gain proportional to the absolute value of e) to a dead zone (gain to small devi-

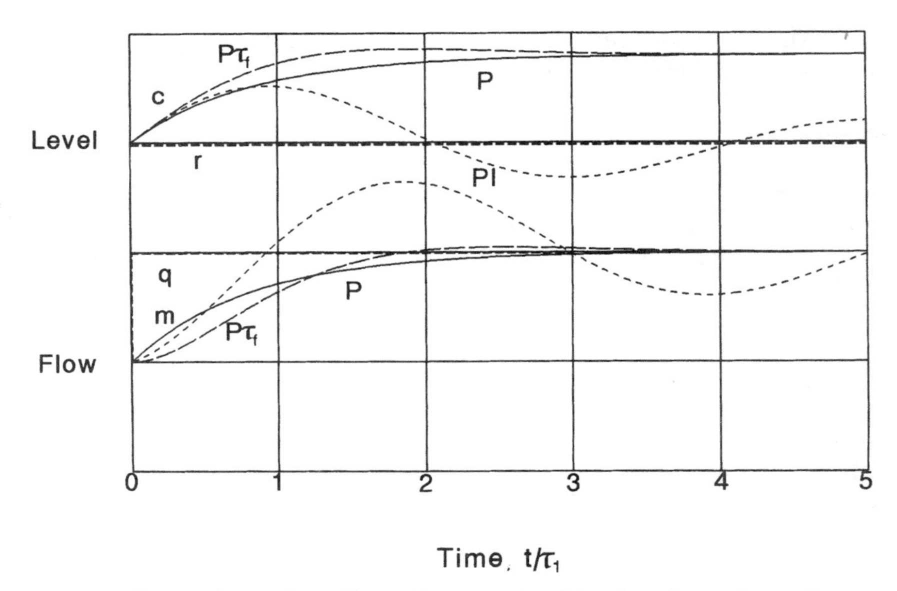

Figure 3.1 Integrating action with a wide proportional band produces slow cycling in a level loop, whereas proportional control is stable, even with filtering.

ations may be low or even zero), but all have the same basic problem. Because of low proportional gain about the set point, damping is light there, and consequently, the level does not remain in this region. Instead, it tends to cycle between the higher-gain regions on both sides, not settling out between disturbances.

By contrast, proportional control can accomplish the flow-smoothing objective without sacrificing stability—there will be no cycles or even overshoot. To maximize smoothing, P has to be set as high as possible to satisfy the second objective of avoiding overflow and emptying, but the limit is 100 percent. To provide 10 percent margin at both ends of the tank, P should be set at 80 percent. The set point r and bias b could both be set a 50 percent, but there is no magic in this number.

Operators tend to object to wide-band proportional control such as we have here because offset will always be present, as shown in Fig. 3.1 for 80 percent proportional band. Yet the full capacity of the tank cannot be used to provide smoothing unless offset is allowed to exist. Furthermore, it makes sense to operate the feed tank nearly full when running at high production rates and nearly empty when running at low rates or starting up. Because the objective of this loop is so much different from that of most of the loops that operators encounter, it should be treated differently. It makes more sense to set the bias at zero and the set point at the low-level limit, e.g., 10 percent. In this way, the set point will be recognized as a limit rather than as an objective to be achieved.

Alternatively, the set point can be eliminated altogether, since it creates the false impression that it will be followed, with control achieved by the simple linear function

$$m = \frac{c - c_l}{c_h - c_l} \tag{3.9}$$

where c_h and c_l represent the high and low limits of tank level at which the manipulated flow should be 100 and 0 percent. These limits replace the set point and should change the expectations of the operator; controller behavior only needs to be checked in reference to these two limits. To provide the same function using a true proportional controller, the set point would be c_l, the proportional band $c_h - c_l$, and the bias zero.

Under proportional control, outflow responds to inflow as a first-order lag having a closed-loop time constant of $\tau_1(100/P)$, where $P = c_h - c_l$, expressed in percent. The highest rate of change of the manipulated flow occurs at the beginning of the exponential curve for the proportional controller in Fig. 3.1. This initial rate can be reduced markedly by adding a first-order filter either at the input or output of the controller, whichever is more convenient; this also removes high-frequency noise from the controller output. The third pair of curves in Fig. 3.1, labeled $P\tau_f$, were produced by a proportional band of 80 percent and a filter set at 0.4 τ_1. The loop has a damping factor $\zeta = 0.707$, producing an overshoot of 5 percent in both c and m. Even with filtering, however, this loop is more stable than with a PI controller. (The equations describing this second-order closed loop are presented under two-capacity processes below.)

Proportional Plus Derivative Control

Derivative action is not used alone, but always in conjunction with proportional action, in the form of a first-order lead. As such, it is the inverse of a first-order lag, having a phase angle of the opposite sign; when applied to a loop containing a lag—particularly a secondary lag—it can virtually cancel the effect of that lag. This can have a dramatic effect on the behavior of second-order systems without deadtime, as will be demonstrated below.

Derivative action is also valuable in a general sense, providing phase lead to offset the phase lag produced by integration. In a PID controller, then, it is useful in shortening the period of the loop and thereby hastening its recovery from disturbances. In this section, the general features of derivative compensation are presented, along with specific applications that can benefit greatly from PD control.

Derivative of the controlled variable

Since the purpose of derivative action is to add phase lead to the control loop, there is no need to apply it to set point of the controller, which is outside the loop. Furthermore, set-point changes are usually introduced stepwise, and applying derivative action to a step produces a large pulse, overdriving the controller output and causing a large overshoot. Consequently, the preferred algorithm for PD control takes the derivative of the controlled variable only, i.e., not the deviation (which would include the set point):

$$m = \pm\frac{100}{P}\left(r - c - D\frac{dc}{dt}\right) + b \tag{3.10}$$

where D is the derivative time constant.

The deviation and the derivative above cannot be added algebraically, since the derivative leads by 90 degrees of phase. Vector addition must be performed, after which the proportional gain is applied to the resultant—this is the "standard" manner of combining control modes, according to the Instrument Society of America (ISA).[2] However, some controllers are based on an "independent" algorithm, where the proportional gain is *not* applied to the derivative term. This can cause some confusion, especially in the rules used to tune the controller. For example, in applying a PD controller to a two-capacity process below, a relationship is developed between the required derivative setting and the process secondary time constant. This relationship holds only if the proportional gain multiplies both terms as in Eq. 3.10. If it only multiplies deviation and not the derivative term, then changes in the proportional gain will affect the derivative action by changing the phase angle of the controller. Its effective derivative time in terms of the standard controller is then $DP/100$. More attention is directed to phase angle later.

Pneumatic controllers having but one stage of amplification (pneumatic relay) provide derivative action through a lag in the proportional feedback loop within the controller. Some electronic and digital controllers also may be manufactured in this configuration. In this position, derivative action is applied to the controller output rather than the input. As such, it responds to both set point and controlled variable, but only if the output is not saturated—while the output is saturated, it does not respond to either. This behavior produces a very different set-point response from that achieved with derivative action directed toward either the deviation or the controlled variable.

Derivative filtering

Pure derivative action is not attainable. Being the inverse of a lag, a pure lead must have a dynamic gain that increases indefinitely with frequency, just as a lag has a dynamic gain that varies inversely with frequency. This response is unattainable with physical components and undesirable as well, because it would amplify noise to unacceptable levels. Derivative action is therefore always accompanied by a filter of some kind—the lead always accompanied by a lag.

Filtering is actually applied to the proportional as well as the derivative term; the resulting PD controller can be expressed as a lead-lag transfer function:

$$\frac{m(s)}{c(s)} = \frac{100}{P}\left(\frac{1 + Ds}{1 + \alpha Ds}\right) \tag{3.11}$$

where s is the LaPlace operator, and αD is the time constant of the first-order derivative filter. If filtering had been applied only to the derivative term, the effective value of the derivative time constant would be $(1 + \alpha)D$ rather than D.

The maximum dynamic gain of the derivative function is $1/\alpha$ and is often described as the derivative gain limit K_D. It is typically fixed at about 10, although it may be adjustable in some controllers. The derivative gain limit changes not only with α, however, but also as a result of interaction with the integral mode in an interacting PID controller and with the scan period in a digital controller. These tendencies are covered under their respective headings.

With K_D fixed at 10, ref. 3 indicates that the PD controller produces a maximum phase lead of 55 degrees, reached when the period of the loop is $2D$. If K_D were increased to 20, the maximum phase lead would be 65 degrees when the period is $1.4D$. Some improvement in control performance is observed when K_D is raised from 10 to 20, but much more performance is lost when it decreases below 10. The effect of derivative gain on performance is evaluated below under PID control, where integrated-error results may be compared and in the next chapter as a function of controller scan period.

There is a curious device known as the *inverse derivative unit,*[4] which is a lead-lag compensator of the same type, but it has K_D set below 1.0. This makes the lag dominant, and a phase lag results. The original purpose of this device was to stabilize flow loops, where the proportional gain of the PI flow controller could not be reduced

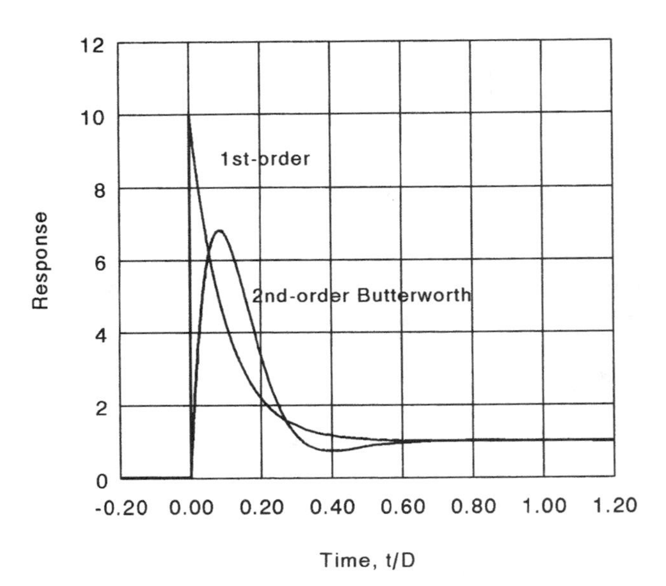

Figure 3.2 The response of PD controllers to a unit step in c with first- and second-order filters, a derivative gain limit of 10, and proportional band of 100.

enough to provide stability due to physical limitations. With this unit in series with the controller, the proportional gain was reduced by K_D. The only other use it might have in process control is as a lead-lag compensator in feedforward systems, or as a set-point filter.

The first-order filter described above leaves the high-frequency gain of the PD controller at $100K_D/P$, which may still amplify noise to an unacceptable level. To apply further gain reduction at high frequencies, a second-order filter must be used. One that is quite effective is the Butterworth filter. It adds 0.5 $(\alpha Ds)^2$ to the denominator of Eq. 3.11. The step responses of a PD controller with the two filters are compared in Fig. 3.2 to indicate the kind of results that may be observed during a bench test. The two pulses have identical areas; the first-order filter gives 63.2 percent recovery at a time equal to αD, but the Butterworth filter gives better noise rejection. The integrated difference between the output transients in Fig. 3.2 and the input step is the size of the step multiplied by the difference between the lead and first-order lag time constants. That difference for both the filters is $(1 - \alpha)D$. Where a second-order derivative filter is not used, there will normally be a separate first-order filter of fixed time constant (not a function of D) applied to all control modes.

Two-capacity processes

There exists an interesting class of processes having little or no deadtime but with a pronounced second capacity. This secondary lag could be in the measuring system or in the manipulated-variable path. An example of the former case is a liquid-level measurement located in a side capacity on a vessel, such as a stilling well or displacer cage. The level in the vessel itself must change to start flow into or out of the well or cage before the level can change there. An example of the latter case is a heated cubicle where the heating element is the secondary lag and the cubicle and its contents represent the primary lag.

The two-capacity process only *approaches* a stability limit under proportional control as the proportional band approaches zero. The closed loop may be modeled following the same procedure used for the integrating process with the PI controller in Eqs. 3.3 through 3.8, assuming an integrating process here as well. The undamped natural period of the loop varies directly with the proportional band, as before:

$$\tau_n = 2\pi\sqrt{\tau_1 \tau_2 P/100} \tag{3.12}$$

where τ_2 is the time constant of the secondary lag. The damping factor, however, now varies *directly* with the proportional band (whereas in the case of the PI controller, the relationship was inverted):

$$\zeta = 0.5\sqrt{\tau_1 P/100\tau_2} \tag{3.13}$$

(Incidentally, this analysis applies equally well to the case of proportional averaging-level control, where a lag is added to the controller to provide further smoothing and noise filtering; τ_2 would represent the filter time constant.)

Figure 3.3 describes the set-point response of a two-capacity non-self-regulating process under proportional control with two different values of proportional band. Although the damping and period are quite different, as expected from Eqs. 3.12 and 3.13, the overshoot in the two cases is similar.

Since proportional control is not capable of eliminating overshoot, another method is needed, and derivative action is the natural choice. The lead term represented by derivative action can effectively cancel the secondary lag, leaving proportional control of a single-capacity process—P can now be set close to zero without developing oscillations, and near-perfect response can be achieved. Because derivative is not a perfect lead, however, the loop will still contain a secondary lag—that of the derivative filter, αD. If D is set equal to τ_2, then the secondary lag will not have been eliminated but simply reduced to a factor of $\alpha\tau_2$. In the simulation of this loop shown in Fig. 3.4, the results are quite satisfactory. The tank level would be prevented from overshoot as measured level c approaches the set point exponentially; P is 2 percent.

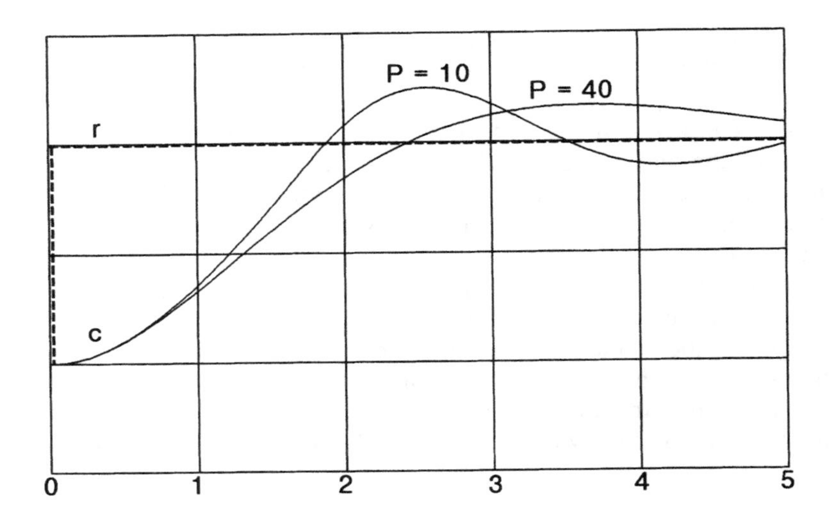

Figure 3.3 Proportional control produces set-point overshoot with a two-capacity process, even when the band is increased several-fold.

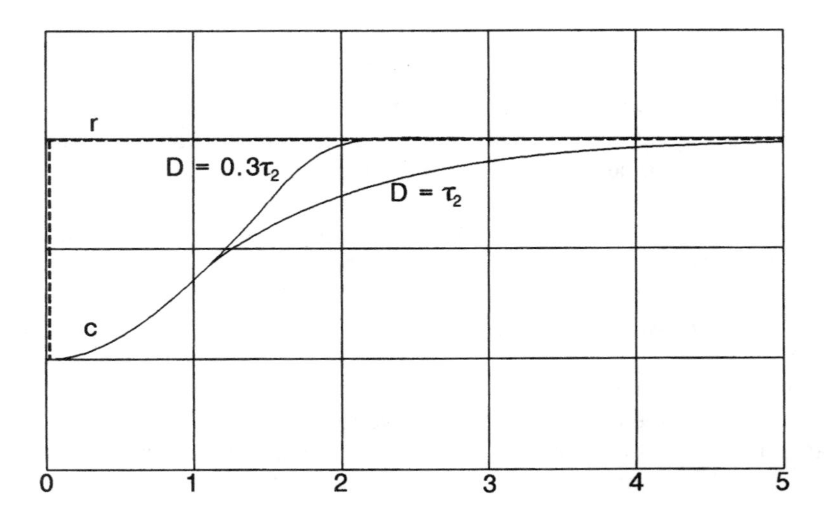

Figure 3.4 PD control avoids set-point overshoot, with the approach trajectory determined by the derivative setting.

In the example of the cubicle with the electric heater, however, where the secondary lag is upstream of the primary lag, c represents the true controlled variable, whose overshoot is to be avoided. A shorter derivative time can now be used, since overshoot in the output of the heater is permissible. Figure 3.4 shows that by setting D to a value of $0.3\tau_2$, a very sharp corner can be turned on the approach to set point. This procedure minimizes the time to reach set point while still avoiding overshoot; again, P is 2 percent.

In both Figs. 3.3 and 3.4, the load is 50 percent. However, two-capacity zero-load processes are also common. An example is the heat treating of batches of metal products described earlier, where the load is nearly zero. Another similar process is the control of batch endpoint, e.g., pH; again, the load is zero, but because of imperfect mixing, proportional control will result in overshoot, which can be eliminated by the addition of derivative action. In a zero-load process, remember that any overshoot is permanent; as a result, D must be set at $(1 + \alpha)\tau_2$ to avoid overshoot. The optimal ratio of D/τ_2 for the two-capacity process is therefore load-dependent, decreasing from $1 + \alpha$ at zero load to 0.3 at 50 percent load and remaining there for further increases. (Remember that this rule applies only to the standard PD controller and must be modified by the proportional gain for use with an independent controller.) The bias must always be matched to the load to avoid offset.

Integral Control

An integral controller positions the manipulated variable as a function of the time integral of the deviation between the controlled variable and set point:

$$m = \pm \frac{1}{I} \int (r - c)\, dt + C \tag{3.14}$$

where the $\pm$ sign again indicates the choice of controller action, I is the integral time constant, and C is the constant of integration.

As with the case of proportional gain, manufacturers are split over the designation of integral action, but probably the majority prefer to identify it as "integral gain" or "reset rate," $1/I$, in "repeats per minute" or "repeats per second." This approach would appear to be consistent with use of proportional gain rather than proportional band, and those manufacturers which use proportional gain also use integral gain. The notion of *repeats* is that the integral of a given deviation will repeat that deviation during the integral time of the controller. For example, an integral controller having a deviation of x percent will move its output x percent during each integral time, its output thereby repeating that deviation each integral time. However, expressing integral in units of time I seems more useful in diagnostic and tuning work because it relates directly to other time functions such as process deadtime and loop period. Furthermore, in a PID controller, I should be adjusted relative to derivative setting D, which is expressed in units of time for all controllers.

The constant of integration C is the combination of all the history of the control action taken to the present, along with the output that the controller was given manually before transfer into the automatic mode. This memory of the past will be quite useful in some instances and troublesome in others, especially where the process conditions have been changed without the controller's participation, i.e., in the open loop. This leads to a condition commonly called *windup*, a subject covered extensively in this book, especially in the last chapter.

Elimination of offset

For any process having a variable and unknown load, integral action is needed if offset is to be eliminated. The ability of the integral control mode to accomplish this end is apparent not so much from its algorithm above as from the derivative of it:

$$\frac{dm}{dt} = \pm\frac{r - c}{I} \tag{3.15}$$

This tells us that the velocity of the output is proportional to deviation and that the output will not come to rest until the deviation is zero. As a consequence, there will be no steady-state offset with integral action in a controller, whether alone or in conjunction with proportional and derivative.

Offset may develop during a ramp in set point, however, which requires a sustained velocity in the output. In process control, ramps always terminate, and offset during a ramp is not usually considered detrimental. The important exception to this is in a digital blending system, where it is desired to have the total flow being delivered through a flowmeter track the desired total while that desired total is changing at a given rate—a moving target. To accomplish this end, a PI controller must be applied to the deviation in total flow while manipulating rate of flow. Functionally, this loop is one of double-integral being necessary to avoid offset during the continuous ramp.

Having eliminated steady-state offset, integrated error now reaches a limit following a disturbance rather than continuing to change indefinitely. The area criteria used to measure controller performance are only useful in the absence of offset. While offset can be eliminated in proportional and PD loops by proper biasing, the use of integral action gives a general relationship between the disturbance and the resulting area by which controller performance can be measured.

Consider the loop in a steady state subject to a disturbance requiring a change in m to reach a new steady state. In moving from the initial to the final value of m, the controller has had to integrate some error. This integrated error can be estimated directly from controller algorithm 3.14, independent of the source of the disturbance or the path taken between the states:

$$\mathrm{IE} = \mp I \Delta m \tag{3.16}$$

The reversal of signs has been introduced to relate IE to the best possible integrated error IE_b, which integrated $c - r$, whereas the controller integrates $r - c$.

If the disturbance were a load change, the change in m would match it; for a set-point disturbance, m would have to change an amount $\Delta r / K_p$. This relationship now gives a means of estimating controller performance based on the magnitude of a given disturbance and the tuning of the controller. To be useful, however, there must be accompanying tuning rules that guarantee stability, which integrated error does not address. Because Eq. 3.16 is not path-sensitive, it contains no information on response features such as overshoot, decay ratio, settling time, or period, all of which are important in identifying loop behavior. In practice, tuning a controller to minimize integrated *absolute* error IAE guarantees stability and minimizes the product of peak deviation and settling time. It also tends to minimize overshoot and decay ratio, which is also generally desirable. Therefore, as a general practice, I tune controllers to minimize IAE; performance then may be reported in terms of IE, with confidence that all other features of the response are acceptable.

The step-load response of a deadtime process under integral control appears in Fig. 3.5, with the controller having been tuned for minimum IAE. The integral time producing this result is 1.6 $K_p \tau_d$, which is also the value of $\mathrm{IE}/\Delta m$ from Eq. 3.16. Best possible IE_b for this same process is $\Delta q K_p \tau_d$. Since $\Delta m = \Delta q$, the performance of this controller on the deadtime process is 1/1.6, or 62.5 percent.

Phase and gain

Passing a sine wave through an integrator produces a cosine wave—the phase angle between input and output is −90 degrees regardless of the period of the wave. This phase lag is the price paid for elimination of offset, and a second integral pays twice the price. While the phase angle is fixed, the dynamic gain vector of the integrator has magnitude G_I varying directly with the period τ_o of a wave passing through and inversely with integral time:

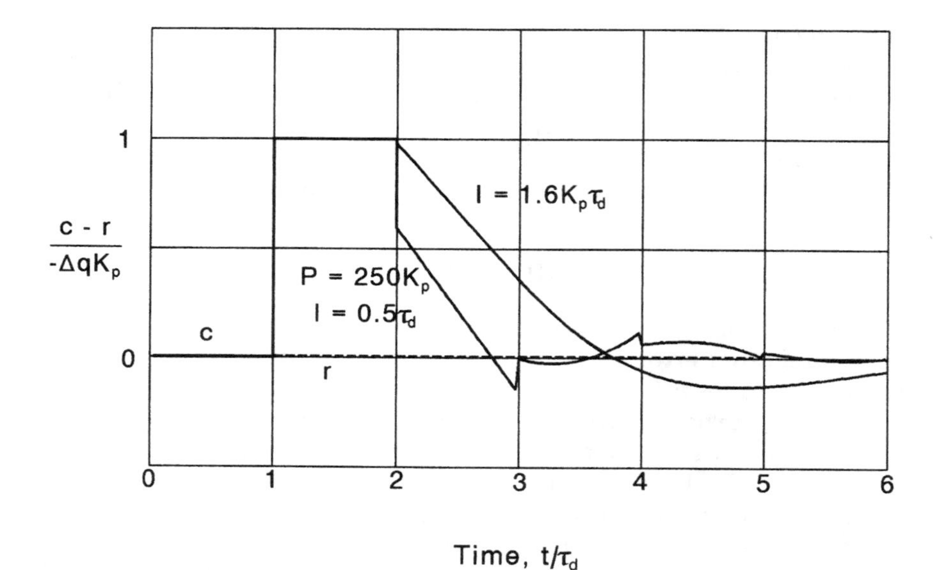

Figure 3.5 Integral and PI control are compared against best possible recovery from a step-load change to a deadtime process.

$$G_I = \frac{\tau_o}{2\pi I} \tag{3.17}$$

The dynamic gain increasing with period verifies the elimination of offset in the steady state (infinite period). In practice, however, the controller gain cannot be infinite in the steady state because there are always limitations. In a pneumatic controller, there is a mechanical gain limit, in an electronic controller a limited amplifier gain, and in a digital controller a limit to numerical precision. But the dynamic gain of the integral controller to fast-moving disturbances is so low that recovery time is extended, producing an integrated error much higher than that associated with PI and PID controllers. This is one reason why integral action is rarely used alone.

An even more compelling reason is the instability produced when attempting to control a non-self-regulating process—connecting two integrators in a loop produces an oscillator. A loop phase of 180 degrees is always present, causing an undamped cycle to develop at the period where the loop gain crosses unity.[5] No amount of tuning can dampen the resulting oscillation—this is simply an unstable combination of controller and process.

Non-self-regulating processes are usually identifiable as inventory control with independent inflow and outflow, liquid-level control being the most common. However, some processes that are normally self-regulating also can exhibit this behavior. For example, a process can lose its self-regulation at zero load where outflow is constant (at zero). A process having a dominant lag and very high steady-state gain also appears to the controller to be non-self-regulating—pH control is a good example of this type. A final example is a chemical process that is normally self-regulating but has a recycle loop added to recover unreacted feed; this recycle can cause components to accumulate where they would otherwise be carried away. In essence, a recycle loop in a plant is a positive-feedback loop capable of canceling the inherent negative feedback characteristic of self-regulation.

Proportional Plus Integral Control

The PI controller combines the stability of proportional control with the offset elimination of integral action to considerable advantage. As with the PD controller, it combines vectors that are 90 degrees apart in phase:

$$m = \pm\frac{100}{P}\left(e + \frac{1}{I}\int e\,dt\right) + C \tag{3.18}$$

According to ISA standards,[2] the two terms are combined before their resultant is multiplied by the proportional gain, whereas in the independent algorithm, proportional gain is applied only to the deviation. Again, this can cause tuning problems by altering the controller's phase angle on a gain change. For example, decreasing the proportional gain increases phase lag and can destabilize a loop as easily as increasing proportional gain.[6] The integral time of the independent controller in terms of the standard controller is $100I/P$.

Another variation of the PI controller replaces e with $-c$ for the proportional term, as was done with the derivative term in the PD controller. This avoids proportional response to a set-point change, which can be helpful in avoiding set-point overshoot after the controller has been tuned for optimal response to load changes. (The tuning dilemma between optimal load and set-point responses is examined in detail in Chap. 6.) The function obtained thereby is equivalent to applying a first-order filter to the set point of the controller in Eq. 3.18, with the filter time constant being equal to the integral time. The integral mode, of course, must always respond to deviation.

Integral time is used here again, although repeats per unit time may be the more common identifier of integral action. Where the integral controller repeated the deviation in the integral time, the standard PI controller repeats the proportional component of output. This distinction is necessary because of the multiplication of both terms by the proportional gain. With an independent algorithm, however, there is no difference from the integral controller.

Phase and gain

The dynamic gain of the PI controller to the controlled variable is the resultant of the proportional and integral vectors. The proportional vector is simply magnitude 1.0, phase angle 0; the integral vector is that of the integral controller described by Eq. 3.17. The resultant is

$$\phi_{PI} = -\tan^{-1}\left(\frac{\tau_o}{2\pi I}\right) \qquad G_{PI} = 1/\cos\phi_{PI} \tag{3.19}$$

Again, proportional gain is normally applied to the resultant; however, the independent algorithm applies it to the proportional vector before summation, causing the phase angle to change with proportional gain. In the examples that follow, tuning rules are developed that relate integral time to the process dynamics and proportional band to process steady-state gain. These rules would have to be modified before use with an independent algorithm.

Although my practice in tuning a controller is to minimize IAE (or produce an overshoot and decay ratio characteristic of minimum IAE) rather than pay any attention to phase margin, the preceding expression is useful in diagnosing a badly mistuned controller. If a control loop is observed to be oscillating with relatively little damping, the controller's phase angle may be estimated by entering the observed period and integral setting into Eq. 3.19. Its phase lag should always be less than 45 degrees for deadtime-dominant processes and closer to 13 degrees for non-self-regulating processes; lag-dominant processes will fall in between, depending on the ratio of deadtime to primary time constant. While this may seem like a rather crude diagnostic, it can quickly identify the mistuned level controller by its 80 degree phase lag, exemplified by the PI loop in Fig. 3.1. A similar diagnostic is provided below for PID controllers.

Integrated error

As done with the integral controller, the integrated error between two steady states can be estimated by solving Eq. 3.18 for m at each of the states and subtracting the results. Because e is zero in both steady states, the change in output is related only to its integral:

$$\text{IE} = \mp\frac{PI}{100}\Delta m \tag{3.20}$$

The only reason for the proportional gain to appear in Eq. 3.20 is that it is normally multiplied by the integral term; for the independent algorithm, this is not done, and therefore, Eq. 3.16 applies.

Figure 3.5 compares the step-load response for the PI controller with that achieved by the integral controller. The controller settings which minimize IAE are $P = 250K_p$ and $I = 0.5\tau_d$. With these settings, IE/Δm is $1.25K_p\tau_d$, falling about halfway between that of the integral controller at 1.6 and the best possible at 1.0. Its performance rating is 1/1.25, or 80 percent, compared with the integral controller's 63 percent. (The independent PI algorithm is capable of the same performance but requires an integral time of $1.25K_p\tau_d$.) With this tuning, the PI loop reaches its limit of stability (undamped oscillations) when the process gain increases by 100 percent or when deadtime increases by 138 percent—its robustness is rated at the lower figure, 100 percent. With the optimally tuned integral controller, the loop reaches its limit of stability when either process gain or deadtime increases by 150 percent, which is then its robustness.

Controller structure

Figure 3.6 shows a configuration with which PI control has been implemented by several manufacturers. Integration is accomplished by positive feedback of the controller output through a first-order lag having the integral time constant I. In transfer function form, this appears as

$$m(s) = \pm\frac{100}{P}e(s) + \frac{m(s)}{1 + Is} \tag{3.21}$$

Combining terms,

$$m(s)\left(1 - \frac{1}{1 + Is}\right) = \pm\frac{100}{P}e(s) \tag{3.22}$$

yields

$$\frac{m(s)}{e(s)} = \pm\frac{100}{P}\left(1 + \frac{1}{Is}\right) \tag{3.23}$$

As long as e in Fig. 3.6 is zero, the positive-feedback loop holds its present value of output, bias b and output m being identical. When e

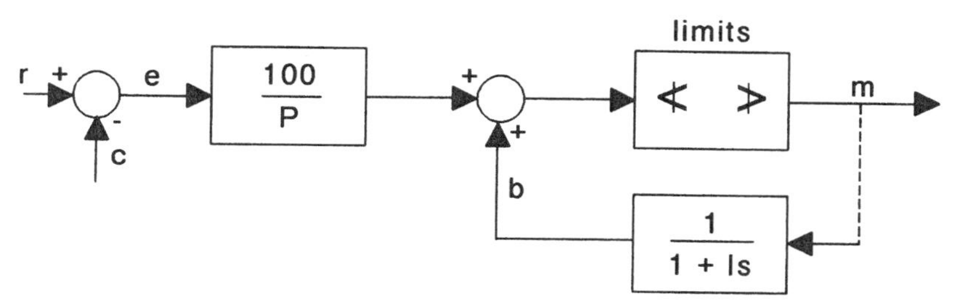

Figure 3.6 Integration can be achieved by positive feedback of the controller output through a first-order lag.

is then stepped to a non-zero value, $100e/P$ is added to b as a proportional change in m. As soon as m differs from b, however, b begins to change at a rate directly proportional to their difference and inversely proportional to integral time. As long as that value of e is maintained, the controller output ramps; should e return to zero, the proportional component of output will be removed and the ramp stopped.

Integral windup

If a deviation is maintained long enough to drive bias b to one of the output limits, the controller is said to be "woundup." In this condition, the controller output will not leave that limit until the deviation reverses sign, i.e., until the controlled variable crosses to the other side of set point. Excessive overshoot is therefore an effective indicator of integral windup. The degree of overshoot caries with the difference between the process load and the controller bias. A controller without limits on its output can windup to full supply pressure or voltage and therefore will require more time to recover than one that has output limits.

The cause of integral windup is an open control loop. The controller output is either at a limit or is unable to affect the controlled variable because of an open circuit, closed block valve, empty vessel, or the like. The load could temporarily exceed the limits of the manipulated variable, or some intervening force such as an operator or competing controller could have opened the loop. Windup protection may be applied in a number of ways, but the condition must first be detected. The usual method of detection is to compare the controller or integrator output with its nearest limit and take corrective action when they are equal. However, in a cascade control system where a primary controller drives the set point of a secondary controller, primary windup begins when the secondary loop opens, independent of the value of the primary output.

Any controller having integral action may windup, but the structure in Fig. 3.6 is particularly easy to protect against it. The dashed connection between the output and the integral lag may be broken, and a signal other than the output may be substituted as integral feedback to stop integration. This procedure effectively converts the PI controller to proportional action with remote bias. The change may be imposed by a switch driven from some logic or by feedback from a process variable indicative of the prevailing load. Several applications are presented in Chap. 9.

PID controllers

The first PID controllers were PI controllers to which derivative action was added by inserting a lag in the proportional feedback loop around the single amplifier. There was no attempt at creating a mathematically pure PID function, as evidenced by the names applied to the control modes. For example, integral action was generally called "reset," and derivative action was "rate." Manufacturers even had their own names for derivative action; Taylor Instrument Companies called it Pre-Act, and The Foxboro Company used Hyper-Reset.

While of simple construction, these controllers are functionally more complex than the more ideal PID controllers developed using electronic and digital technology. They are also quite effective, and some manufacturers continue to produce the same functionality in electronic and digital controllers. Mathematical purity has no particular value in control—performance, robustness, and ease of operation and tuning are the real issues.

The ideal or noninteracting PID controller is a simple linear combination of the three modes:

$$m = \pm\frac{100}{P}\left(e + \frac{1}{I}\int e\,dt - D\frac{dc}{dt}\right) + C \tag{3.24}$$

Figure 3.7 shows one possible structure. Notice that integration and differentiation are parallel operations, consistent with Eq. 3.24. By applying the proportional gain to the deviation and derivative, adjusting its value will produce at worst a transient change in output and

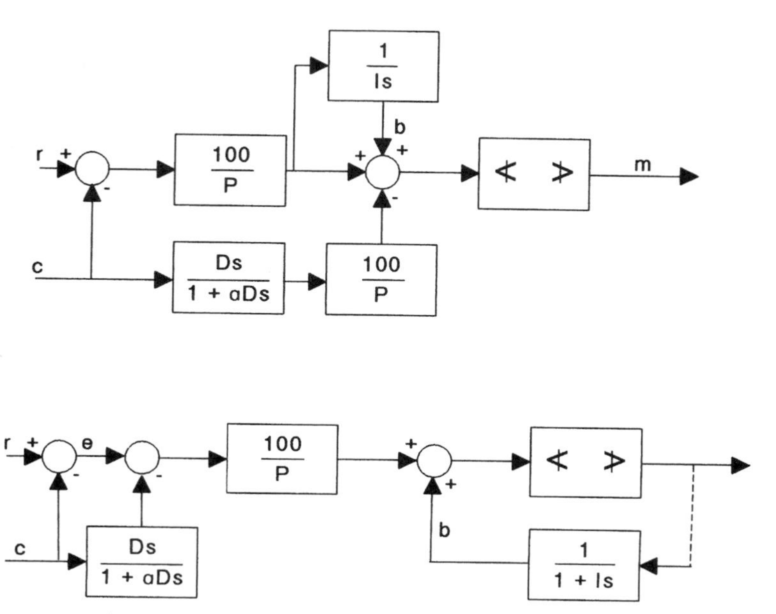

Figure 3.7 A noninteracting PID controller appears above, with an interacting controller below.

no change at all in a steady state. If the proportional gain were to be applied to the output, however, any adjustment would change the constant of integration, producing the equivalent of a step load change. Again, the ISA standard algorithm is shown, where proportional gain is applied to all terms.

In a steady state, the derivative term is always zero. As a result, the integrated error between any two steady states is the same function for the PID controller as for the PI controller, as given in Eq. 3.20. However, derivative action affects the IE of a properly tuned PID controller by allowing a significant reduction in integral time over the PI controller.

Two-stage interacting controllers

The lower diagram in Fig. 3.7 describes a two-stage interacting controller, the second active amplifying stage being used for derivative action. It can best be described by the following transfer function:

$$\frac{m(s)}{c(s)} = \pm\frac{100}{P}\left(\frac{1 + Ds}{1 + \alpha Ds}\right)\left(1 + \frac{1}{Is}\right) \tag{3.25}$$

Multiplying the two factors and separating terms transforms Eq. 3.25 into the equivalent noninteracting controller:

$$\frac{m(s)}{c(s)} = \pm\frac{100}{P}\left(1 + \frac{D}{I}\right)\left[\frac{1}{1 + \alpha Ds} + \frac{1}{(I + D)s} + \frac{s}{(1/D + 1/I)(1 + \alpha Ds)}\right] \tag{3.26}$$

One result of interaction is that proportional gain is increased by the ratio of derivative to integral settings:

$$K_c = \pm\frac{100}{P}\left(1 + \frac{D}{I}\right) \tag{3.27}$$

The effective integral time includes the derivative setting:

$$I_{\text{eff}} = I + D \tag{3.28}$$

and the effective derivative time includes the integral setting:

$$D_{\text{eff}} = \frac{1}{1/D + 1/I} \tag{3.29}$$

Finally, the derivative gain limit is no longer simply $1/\alpha$:

$$K_D = \frac{1 + 1/\alpha}{1 + D/I} \tag{3.30}$$

Within limits, the effective settings of an interacting PID controller can be matched to the settings of a noninteracting controller. These limits are reached when $D = I$, producing $D_{\text{eff}}/I_{\text{eff}}$ of 1/4. It will be demonstrated that minimum IAE control for first-order processes with deadtime requires the ratio of the effective values to be less than 1/4. With second-order processes, however, optimal tuning requires the ratio to increase with the ratio of secondary lag to deadtime, giving the noninteracting controller more of a performance advantage. Even with first-order processes with deadtime, transforming the optimal settings from one controller to the other using Eqs. 3.27 through 3.29 does not produce optimal settings for the other controller. Differences in derivative gain require other adjustments to be made. Comparisons of controller performance follow later in the chapter, along with the mode settings that produced them.

For load changes, the integrated error for this controller may be estimated using the set values of P and I in Eq. 3.20, just as is done for the noninteracting controller. For set-point changes, however, the IE for the interacting controller contains an additional term $D\ \Delta r$, which does not appear with the noninteracting controller. This term has a very significant influence over set-point response and is covered in Chap. 6 under tuning for set-point response.

Single-stage interacting controllers

There are two basic configurations of single-stage controllers, both shown in Fig. 3.8. Both achieve derivative action using a lead-lag function in a negative-feedback loop around an amplifier (e.g., pneumatic relay) having a high gain G. One, exemplified by the Foxboro model 40, 43, and 58 pneumatic controllers, takes integral feedback in *series* with the derivative lead-lag function. The other, exemplified by the Taylor

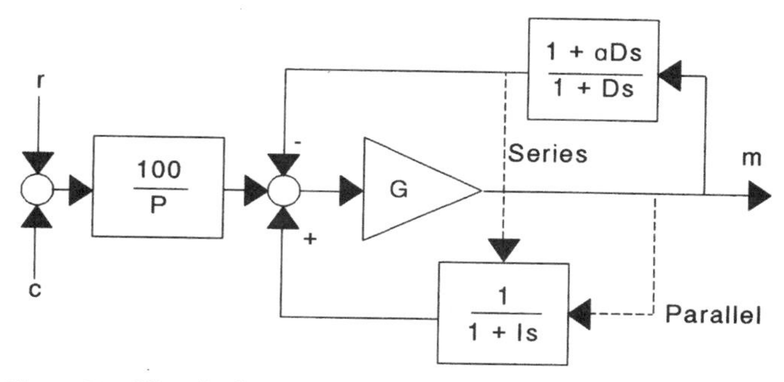

Figure 3.8 The single-stage PID controller may have either serial or parallel feedback loops providing integral and derivative action.

Fulscope controller, has integral feedback taken directly from the controller output and therefore in *parallel* with derivative feedback. This distinction is important, because in the parallel configuration, the positive-feedback loop can be made faster than the negative-feedback loop by setting $I < D$, which actually *reverses* the control action. In the series configuration, the negative-feedback loop always dominates.

The transfer function for the series-connected controller can be derived by summing all the terms at the input to the amplifier:

$$\frac{m(s)}{G} = \pm\frac{100}{P}e(s) - m(s)\left(\frac{1 + \alpha Ds}{1 + Ds}\right)\left(1 - \frac{1}{1 + Is}\right) \approx 0 \quad (3.31)$$

The sum approaches zero because of the high gain of the amplifier. The transfer function in its factored form then appears identical with Eq. 3.25 and therefore produces the same interaction among the control modes.

The controller with parallel feedback has a more complex transfer function:

$$\frac{m(s)}{e(s)} = \pm\frac{100}{P}\left[\frac{1 + Ds}{1 - (1 - \alpha)D/I + \alpha Ds}\right]\left(1 + \frac{1}{Is}\right) \quad (3.32)$$

which rearranges as

$$\frac{m(s)}{e(s)} = \pm\frac{100}{P}\left[\frac{1 + D/I}{1 - (1 - \alpha)D/I + \alpha Ds}\right]\left[1 + \frac{1}{(I + D)s} + \frac{Ds}{1 + D/I}\right] \quad (3.33)$$

The principle difference between the parallel single-stage controller and other interacting controllers is the denominator of its proportional gain:

$$K_c = \pm\frac{100}{P}\left[\frac{1 + D/I}{1 - (1 - \alpha)D/I}\right] \quad (3.34)$$

As D approaches I, the proportional gain of the parallel controller climbs much faster than that of other interacting controllers; at $D = I/(1 - \alpha)$, it approaches infinity, and above that value, the controller action is reversed. This factor has had some impact on history, because Ziegler and Nichols used the Fulscope controller to develop their optimal tuning rules for PID controllers. Their recommended settings for derivative and integral terms placed them in a 1/4 ratio rather than a ratio of 1/2.7 favored by myself on a similar process. The estimate of IE for this controller using Eq. 3.20 must be based on the effective value of I, with $1/K_c$ from Eq. 3.24 substituted for $P/100$.

When windup protection is added to the Foxboro pneumatic controllers mentioned above, the customary series connection of the two time constants cannot be made—any leakage in the external feedback circuit will cause pressure loss across the derivative restrictor and result in offset. As a consequence, these modified controllers are configured with parallel feedback. The early model 40 controller had Bourdon restrictors for integral and derivative settings mounted next to each other, with a link connecting them to prevent setting $D > I$.

Performance and robustness comparisons

The first comparison to be made is between the performance of PI and PID controllers. For a pure deadtime process, derivative control has no contribution to make and is set to zero. Therefore, there is no performance difference between PI and PID controllers on that process. As the ratio of lag to deadtime increases, derivative control offers more advantage, indicated by a growing performance difference between PI and PID controllers. Figure 3.9 compares their responses for a step load change on a first-order non-self-regulating process—a lag-dominant process of the kind used by Ziegler and Nichols to develop their tuning rules.[1] Table 3.1 lists the controller settings used and

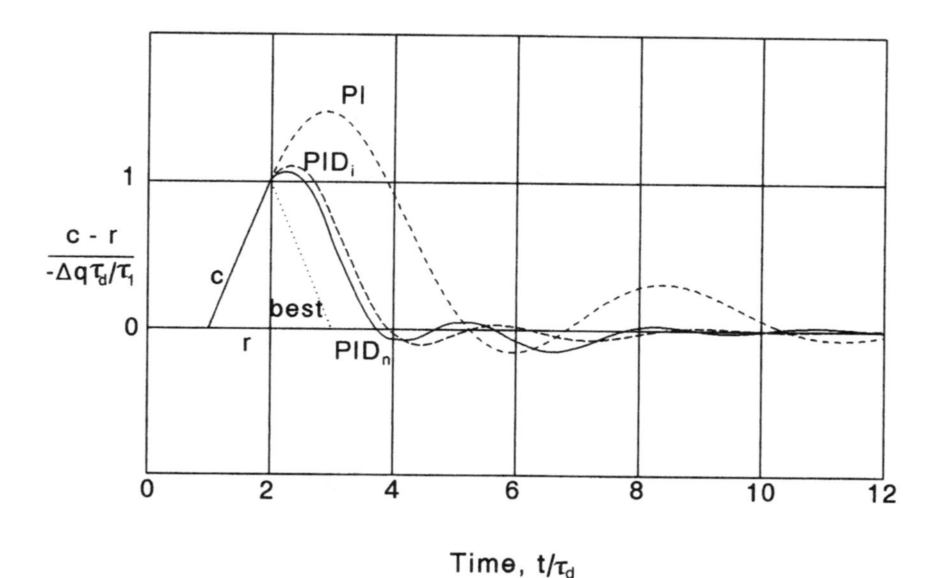

Figure 3.9 PI and PID control are compared against best practical recovery from a step-load change to a first-order non-self-regulating process with deadtime.

TABLE 3.1 Optimal Settings for First-Order Non-Self-Regulating Processes

Controller	$P\tau_1/\tau_d$	I/τ_d	D/τ_d	τ_o/τ_d	IE_b/IE
PI	108	4.00	0	5.45	23%
PID_n	78	1.90	0.46	2.90	68%
PID_i, $\alpha = 0.10$	108	1.60	0.58	3.10	58%
(eff.)	(79)	(2.18)	(0.47)		
PID_i, $\alpha = 0.05$	108	1.48	0.63	3.00	63%
(eff.)	(76)	(2.11)	(0.48)		

the performance achieved.

The reductions in IE and period τ_o contributed by derivative action are quite evident. The natural period for this loop (undamped cycle under proportional control) is $4\tau_d$. The period of about $3\tau_d$ for the PID controllers indicates that derivative action is adding more than enough phase lead to the loop to overcome the phase lag contributed by the integral mode. These controllers both have $\alpha = 0.1$; K_D for the noninteracting controller is therefore 10, whereas that of the other is reduced to 8.07 due to interaction.

Decreasing α to 0.05 raises K_D to 20 for the noninteracting controller but produces no noticeable improvement in performance, nor do the optimal settings change. In the interacting controller, however, minimum IAE requires reducing I and raising D; K_D is raised to 14.7 and performance from 58 to 63 percent.

The robustness of this non-self-regulating process is about the same with all the controllers. Having no steady-state gain, an equivalent stability test can be produced by changing the process dynamic gain inversely with its time constant. With the PI controller, the loop goes unstable on a deadtime increase of 38 percent or a (dynamic) gain increase of 51 percent, so the robustness is 38 percent. With the noninteracting PID controller, the stability limit is reached on a deadtime increase of 35 percent or a gain increase of 30 percent, with a robustness of 30 percent. With the interacting PID controller, the robustness is 35 percent for both process parameters when $\alpha = 0.1$, falling to 32 percent when $\alpha = 0.05$.

A second-order process has the same IE_b as the first-order process yet is more difficult to control. As a result, the three controllers all suffer a decrease in performance. Figure 3.10 and Table 3.2 compare them for a process having equal deadtime and secondary lag; scales and mode settings are based on the sum of deadtime and secondary lag: $\Sigma\tau = \tau_d + \tau_2$. The PI controller performs quite poorly, and the difference between the two PID controllers has widened (both have $\alpha = 0.1$). Observe that the D/I ratio for the noninteracting controller has been reduced to 1/2.5, which is beyond the capability of the interacting controller. The lower performance of the PI controller brings

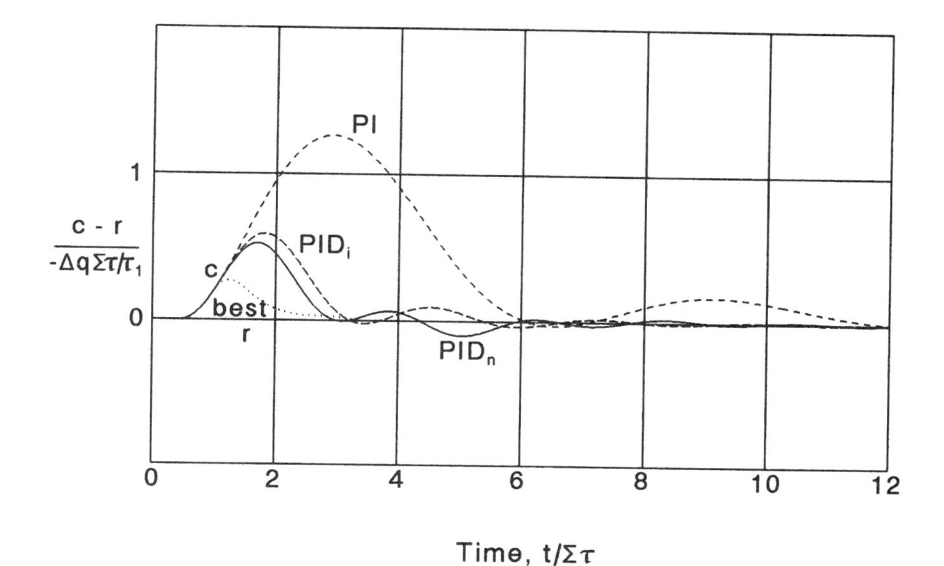

Figure 3.10 The PID controllers are both more effective when half the deadtime of the process in the previous figure is replaced with an equal secondary lag.

TABLE 3.2 Optimal Settings for Second-Order Non-Self-Regulating Processes

Controller	$P\tau_1/\Sigma\tau$	$I/\Sigma\tau$	$D/\Sigma\tau$	$\tau_o/\Sigma\tau$	IE_b/IE
PI	180	3.70	0	5.55	7.5%
PID_n	86	1.38	0.55	2.25	42%
PID_i	128	1.38	0.66	2.63	28%
(eff.)	(86)	(2.04)	(0.49)		

with it an improvement in robustness up to 70 percent, but the robustness of neither PID loop is affected by the secondary lag.

Notation

b Output bias

c Controlled variable

c_h High limit of c

c_l Low limit of c

C Constant of integration

d Differential operator

D Derivative time

e Deviation

G Gain
h Output of first capacity
I Integral time
IE Integrated error
IE_b Best practical integrated error
IAE Integrated absolute error
K_c Proportional gain of a controller
K_D Derivative dynamic gain limit
K_p Process steady-state gain
m Manipulated variable
q Load variable
r Set point
s LaPlace operator d/dt
t Time
x Variable
α Ratio of derivative filter time to derivative time
τ_d Deadtime
τ_n Natural period
τ_o Period
τ_1 Primary time constant
τ_2 Secondary time constant
ϕ Phase angle

Subscripts

eff Effective
I Integral
PI Proportional-integral

References

1. Ziegler, J. G., and Nichols, N. B., "Optimum Settings for Automatic Controllers," *Trans. ASME,* November 1942, pp. 759–768.
2. *Process Instrumentation Terminology* (ISA-S51.1), Instrument Society of America, Research Triangle Park, N.C., 1979, p. 11.
3. Shinskey, F. G., *Process Control Systems,* 3d Ed., McGraw-Hill, New York, 1988, p. 135.
4. Moore, C. D., "The Inverse Derivative—A New Mode of Automatic Control," *Instruments,* March 1949, pp. 216–219.
5. Shinskey, *op. cit.,* p. 25.
6. Gerry J. P., "A Comparison of PID Control Algorithms, *Control Eng.,* March 1987, pp. 102–105.

Chapter

4

Sampling Effects

Most controllers are now digital in nature, operating on a cyclic basis rather than continuously as do analog controllers. Consequently, the dynamic element of sampling plays a major role in most control loops and is often not given enough consideration. It affects performance, robustness, and tuning as well. Digital controllers have other limitations, too, in resolution and communication which also can degrade control and are covered in this chapter.

The dynamic response of the sampling element is examined first independently of the controller. This allows the investigation of the behavior of loops containing a sampled measurement and a continuous controller or a controller sampled at a much faster rate than the measurement. This insight is useful in itself but also helps predict the behavior of the sampled controllers covered later.

Controlling Sampled Variables

The classic example of a sampled measurement is the chromatographic analyzer. It withdraws a sample of process fluid, chasing it through a packed column with a light carrier gas. The sample is fractionated by the column, with its various fractions emerging at different times. Appearance of the fractions at the exit is measured by their difference from the carrier in some physical property such as thermal conductivity or density. Components can be identified by the time required for them to emerge, and their concentration in the sample can be determined by the integration of the measured physical-property difference over time. When the components of interest have been measured, they are reported, and the analyzer is backflushed and purged before the next sample is taken.

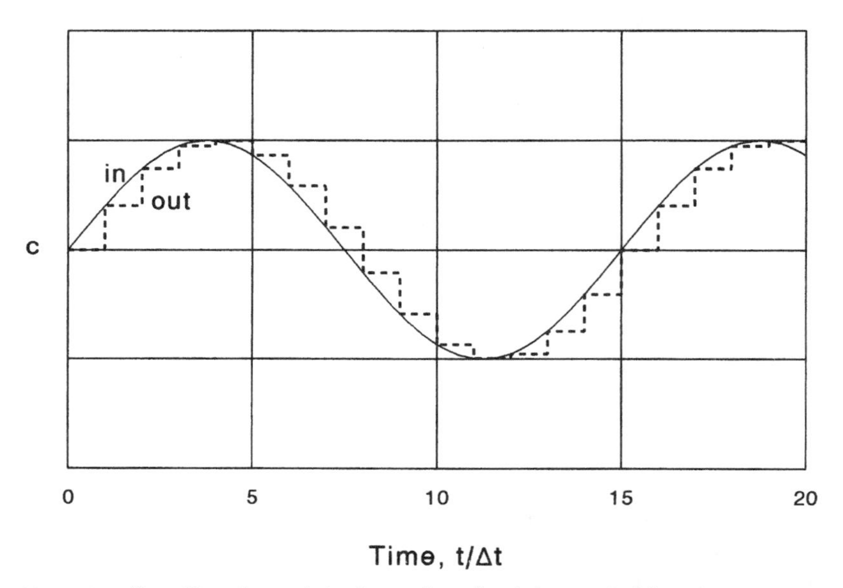

Figure 4.1 Sampling always introduces phase lag into a control loop.

The interval from taking the sample until its composition is reported is deadtime, which affects the control loop exactly as deadtime in a continuous loop. Additionally, however, the control loop is open between samples, which introduces more phase lag. Cycle times for chromatographic analyzers vary from about 2 minutes for separating light hydrocarbons to 20 minutes or more for separating heavier or more complex mixtures. It is also common to multiplex these expensive analyzers among two or more sample points, which increases the sample interval of a control loop without adding to the deadtime.

Neglecting the deadtime, simply sampling a continuous sine wave as shown in Fig. 4.1 produces a staircase that parallels the sine wave but is delayed, on average, for one-half sample interval. Each sample is "held" until the next sample appears, i.e., for one sample interval Δt. This method of holding the last result is specifically called a *zero-order hold* and is used almost exclusively in industry. A *first-order hold* would move from sample to sample in a series of ramps by connecting the sample points with straight lines instead of steps. It amounts to averaging the values of the last two samples. While avoiding the steps, the averaging process doubles the phase lag of the sampler.

Phase lag of sampling

As soon as a sample is taken from the sine wave in Fig. 4.1, it has the same value as the point on the wave and is fully up to date. However, the longer it is held, the older it becomes until it is updated at the end of the sample interval. Its average age over time is therefore half the sample interval. Not surprisingly, its phase lag is half that of an equivalent deadtime:

$$\phi_\Delta = -180°(\Delta t/\tau_o) \tag{4.1}$$

where τ_o is the period of the sine wave. Recognize that the phase lag produced is half as much as that of an equal value of deadtime, because, on average, the samples are one-half sample interval old.

This relationship holds quite well for phase lags to 45 degrees, which covers those loops where the sampling element does not dominate the dynamic response. As the sample interval approaches and exceeds half the period, the original sine wave cannot be reconstructed from the samples. This property is called *aliasing,* because more than one sine wave can produce the same sampled result. While of importance in communications, this is not of much significance in control loops because it can only happen where the sampling element shifts phase to −90 degrees and farther. In most control loops, there are other dynamic elements which typically produce more phase lag than the sampler, keeping its contribution low. However, even in those loops where the sampling element dominates, considerations other than phase lag apply, since the response tends to be very nonsinusoidal; this situation is covered in the next section. In any case, increasing the sample interval in a closed loop tends to increase the period of the loop as well, so the aliasing problem is not an important consideration here.

If a first-order hold were used, it would connect all the sample points together at the trailing edges of the steps in Fig. 4.1, producing twice the phase lag indicated by Eq. 4.1, equivalent to the same value of deadtime. The information presented by the sampler becomes a full sample interval old. The same result obtains if the average value of the variable over the sample interval is passed to the controller at the end of each sample interval. This digital averaging technique is common to some analog-digital converters.

Because deadtime (or effective deadtime in a multilag process) determines the period of oscillation of a control loop, the phase contribution of the sampling element can be related to it. By contributing half as much phase lag, sampling extends the period of the loop half as much as an equivalent increase in deadtime. Given a process deadtime τ_d, adding a sampling interval Δt will cause the period of the loop to increase by the factor

$$f(\Delta t) = 1 + \Delta t/2\tau_d \tag{4.2}$$

For example, suppose a process has a deadtime of 1 minute and cycles at a period of 4 minutes under proportional control. Adding a sample interval of 0.5 minute gives $f(\Delta t) = 1.25$, and the period of the loop can therefore be expected to increase by 25 percent to 5 minutes. The sampler would be producing a phase lag of 18 degrees in this loop. Recognize, however, that using a first-order hold or digital averaging doubles the phase lag of the sampler, effectively removing the 2 in the denominator of Eq. 4.2.

If the process in question were deadtime-dominant, its dynamic gain would not be a function of period. In this case, the only impact of the sample element would be to augment the period, the controller's dynamic settings, and also the integrated error by $f(\Delta t)$. The dynamic gain of lag-dominant and non-self-regulating processes, however, varies directly with period. For these processes, the proportional band also must be increased with $f(\Delta t)$, with the result that integrated error increases with $f(\Delta t)^2$.

Figure 4.2 compares the step load responses of two PI controllers on a lag-dominant process; one controller samples twice each deadtime; the other, 10 times as fast. The $f(\Delta t)$ factors for the two controllers are 1.25 and 1.025, respectively, whose ratio is 1.22. With each controller tuned for minimum IAE, the integrated error produced by the step change in load is 57 percent higher for the slower controller, although 1.22^2 represents an increase of only 49 percent. Apparently, the effect of the sampling goes beyond $f(\Delta t)$, since using it underestimates the integrated error. The impact of sampling is

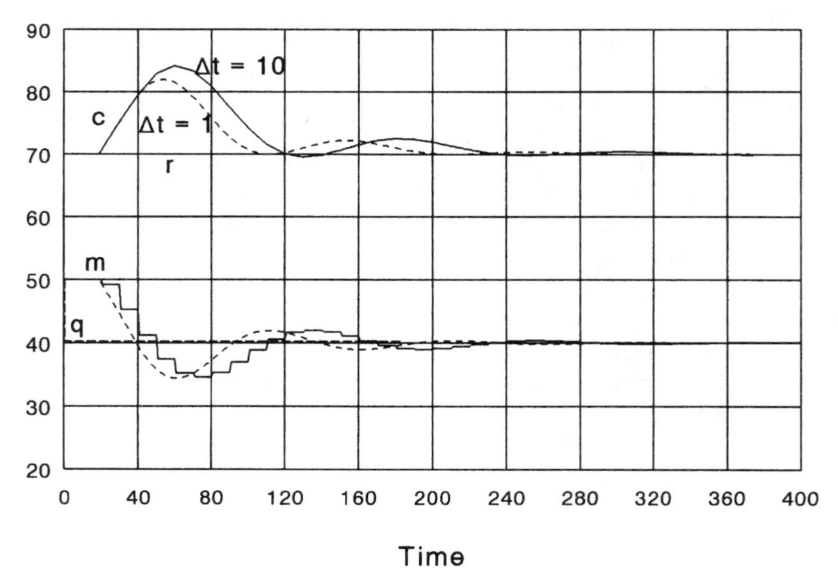

Figure 4.2 Increasing the sample interval of a PI controller produces a larger and longer deviation in response to a step load change to a lag-dominant process.

even worse when derivative action is applied, but this subject is deferred until later in the chapter.

Digital filtering

Filtering is commonly applied to the sampled measurement both for noise rejection and to smooth the abruptness of the sampler output. A simple filter would average the last two samples, as described earlier, causing the filtered output to change only half as much between samples as the input. Integration of the variable over the last sample interval produces a similar result. The dynamic gain is thereby reduced to half the steady-state gain, but a delay of half the sample interval is also added. This doubles the phase lag of the sampler to that of an equal deadtime, as described above. This method of filtering is actually quite common in analog-digital converters and may be applied without the user being aware of its presence. The appropriate phase penalty needs to be assessed by removing the 2 from the denominator of the $f(\Delta t)$ factor in Eq. 4.2. Following the same logic, averaging the last five samples reduces the dynamic gain to 0.2 but adds a delay of 2.5 sample intervals, resulting in a total effective deadtime of 3 sample intervals.

Exponential filtering is more effective than simple averaging in that recent samples are given more weight than older ones, producing less delay. The process is simpler, too, in that only one signal (the filter output) needs to be carried from one sample to the next. A first-order digital filter is derived from the equation describing a first-order lag:

$$\frac{dy}{dt} = \frac{x - y}{\tau} \tag{4.3}$$

where x and y are the filter input and output, respectively, and τ is its time constant. The output can be updated each sample interval as follows:

$$y = y + (x - y)\,\Delta t/\tau \tag{4.4}$$

where the new value of y on the left is found using its last value on the right and the new value of x. The ratio $\Delta t/\tau$ is a *forward-difference operator,* in that it updates y based on the difference between the new value of input x and the last value of output y. Equation 4.4 becomes less representative of Eq. 4.3 as τ decreases, and it produces a "divide by zero" error when τ is set to zero. Filtering is turned off when $\tau = \Delta t$ rather than zero, which is confusing at best. This is actually the low limit for τ, for lower values will cause oscillation, which expands below $\tau = \Delta t/2$.

The more common approach to digital filtering uses a *backward-difference operator,* which is based on a comparison of the new value of x and the updated value of y:

$$\Delta y = (x - y - \Delta y)\,\Delta t/\tau \tag{4.5}$$

which, when solved, produces

$$y = y + (x - y)\frac{\Delta t}{\Delta t + \tau} \tag{4.6}$$

The backward-difference operator is the last term on the right. Setting τ to zero now turns filtering off, which is the expected result.

The true response of a first-order lag to the step changes produced by the sampler is exponential:

$$y = y + (x - y)\,(1 - e^{-\Delta t/\tau}) \tag{4.7}$$

the *exponential* operator being the last term in the equation.

Table 4.1 compares the three operators for various ratios of $\Delta t/\tau$. For ratios of 0.1 or less, the forward-difference operator is closer to the exponential than is the backward-difference operator, although the errors of both in that range are relatively small. There is still another operator called the *Tustin,* which averages the forward and backward differences; it is more accurate than the backward-difference operator for ratios of 1 or less but far worse as the ratio increases toward infinity. Most PID controllers use the backward-difference operator, while some special-purpose controllers such as the Dahlin use the exponential operator.

Absolute accuracy or mathematical purity is not really as important as stability and ease of use, which is why the backward-difference operator is used commonly in controllers. The differences among the various operators principally affect the rules used to tune the con-

TABLE 4.1 Comparison of the Difference Operators

$\Delta t/\tau$	$\Delta t/(\Delta t + \tau)$	$1 - e^{-\Delta t/\tau}$
0.00	0.00	0.00
0.01	0.0099	0.0100
0.10	0.0909	0.0952
0.50	0.3333	0.3935
1.00	0.5000	0.6321
2.00	0.6667	0.8647
10.00	0.9091	1.0000
∞	1.0000	1.0000

troller and the estimate of integrated error based on its settings. For example, the true integral time for a controller using a backward-difference operator is really $I + \Delta t$, as has been determined by error integration during closed-loop testing. Using the forward-difference operator, its true value is I, although it must be limited above zero.

If a second-order filter is needed, the Butterworth filter is recommended. It requires an intermediate variable w to be updated, here using the backward-difference operator:

$$w = w + (x - w - y)\frac{\Delta t}{\Delta t + 0.5\tau} \tag{4.8}$$

followed by another first-order equation:

$$y = y + w\frac{\Delta t}{\Delta t + \tau} \tag{4.9}$$

All these filters use the present value of input to produce a new output and therefore will have a more favorable phase characteristic than one that uses a previous value of input.

Controlling sample-dominant processes

Consider the control of a process whose dynamics are inherently faster than the sampler such that they are masked by it. An example would be a blending system whose product is sampled periodically by an analyzer. Following a step load change as shown in Fig. 4.3, the true composition of the product may reach a steady state before the next sample is taken. Then the deviation appears at the controller input as a step, which is held until the next sample.

Using proportional plus integral action, it is possible to drive the true composition back so that it crosses the set point precisely at the time the next sample is taken. While it may not remain there, if the controller is properly tuned, the true composition will return to the set point before the third sample is taken, requiring no further action from the controller. In this *deadbeat* response, as described in Fig. 4.3, the controlled variable is only away from the set point for one sample interval. Given the length of the sample interval, no better load response can be achieved.

To accomplish this response requires a very specific combination of proportional and integral settings:

$$I = \tau_d + \tau_1 \qquad P = 100K_p\,\Delta t/I \tag{4.10}$$

The process in Fig. 4.3 has a gain K_p of 1.0, and its deadtime τ_d and lag τ_1 are 10 and 5, respectively, compared with a sample interval of 40. The proportional band is therefore 267 percent, and integral time

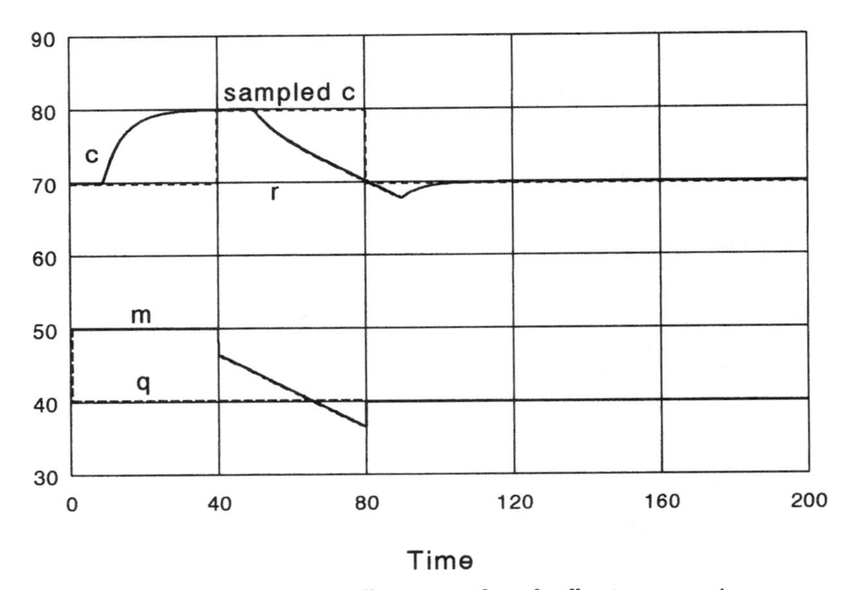

Figure 4.3 A continuous PI controller can produce deadbeat response in a process dominated by a sampler.

is 15. The proportional step caused by the first sample is removed by the second but is needed to force the true composition to cross the set point and return. The *PI* product causes the steady-state change in output m to match the change in load q; the load is not measured, however, but inferred from the change in sampled variable c and the estimated value of K_p. This is the only combination P and I that will produce deadbeat response in this loop.

If the proportional band is lower than optimal, overshoot will result, followed by a damped cycle having a period of two sample intervals. Control is robust, however, in that P would have to be reduced to half it optimal value to produce an undamped cycle. Recognize that although the sample interval is half the period, there can be no aliasing error, for that would reduce the loop gain and dampen the oscillation. In other words, aliasing is not likely to be present in a cycling control loop because it is a stabilizing influence.

Control of sampled lag-dominant processes

The last process was relatively predictable because it could reach a steady state between samples, allowing an estimate of the load change. If the process time constant exceeds the sample interval, which is the more common case, the predictability is lost, and the loop behaves more

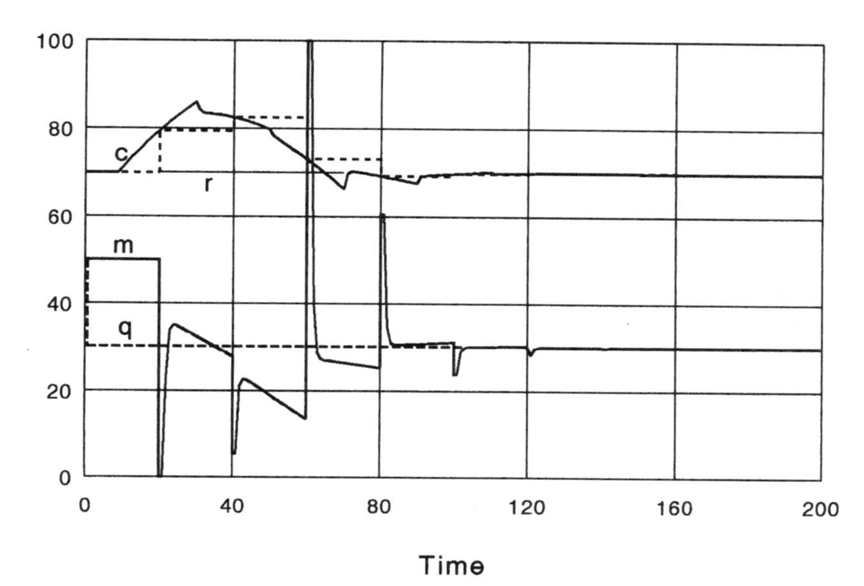

Figure 4.4 Using derivative action on a sampled measurement produces unacceptable spikes.

like other lag-dominant loops, but with the additional phase shift of the sampler. In this light, PID control could be applied, except for one obstacle.

Consider a typical lag-dominant process such as a distillation column, whose product composition is measured by a chromatographic analyzer. Because sampling changes the controlled variable stepwise, derivative action responds by producing large spikes, as shown in Fig. 4.4. The amplitude of the spikes is equal to the size of the steps in c multiplied by $100/P\alpha$, where α is the ratio of the derivative filter time to the derivative time, typically 0.1. Therefore, with a proportional band close to 100 percent, a 3 percent step input produces a 30 percent spike. While the controller is effective in returning the sampled variable back to set point, these large transients in the manipulated variable are unacceptable for most applications. Full-scale swings in either reflux or boilup in a distillation column can cause flooding in the upward direction and empty trays in the downward direction.

As a consequence, derivative control is usually turned off in chromatographic analyzer loops. This is unfortunate, because it has been shown to reduce IE by a factor of 2 compared with PI control on other lag-dominant processes. The difference is not quite so dramatic as this application, because some of the effectiveness of derivative action is blunted by saturation of the controller output by the first spikes. The process being controlled in Fig. 4.4 has $K_p = 2$, $\tau_d = 10$, and $\tau_1 = 40$, with $\Delta t = 20$; minimum IAE controller settings are $P = 94$, $I = 22.7$, and $D = 10.2$, giving an IAE/Δm = 25.6. Removing derivative action and retuning for minimum IAE resulted in the same P, with I increased to 42, raising IAE/Δm to 42.4, an increase of 65 percent.

Rather than forsaking derivative action altogether, a filter can be added to remove the objectional spikes while saving most of the control effort. However, no more filtering should be applied than necessary to achieve that end, because filtering works at cross-purposes with control by trying to keep the controller output from changing. An analog filter applied to the output of the sampler or a digital filter operating at a scan period 10 to 20 times shorter than the analyzer's sample interval is required. (The filter should not be applied to the output, since information is lost during saturation.) Figure 4.5 shows the results of filtering and retuning all three modes for minimum IAE. Using a filter time of 4 required increasing P to 110, I to 25.7, and D to 11.6, which raised IAE/Δm to 30.7, 20 percent higher than the unfiltered PID controller but much lower than the PI controller. Accommodating the filter into the tuning rules for the controller is quite simple; the effective deadtime of the loop is taken to be the sum $\tau_d + \Delta t/2 + \tau_f$, where the last term is the filter time.

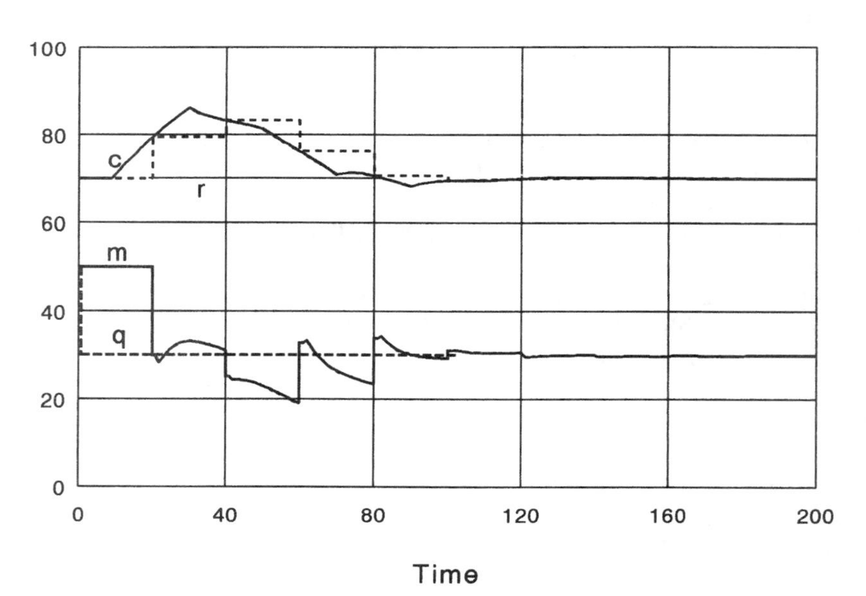

Figure 4.5 A first-order filter having a time constant of $\Delta t/5$ is enough to eliminate the spikes.

Synchronized sampling control

One more method exists for retaining derivative action on a sampled controlled variable while avoiding spikes. It involves synchronizing the controller to the sampled measurement for one calculation or a short period of automatic operation and then leaving it in manual for the duration of the sample interval. A signal from the analyzer must first communicate that a new result has arrived—then the controller takes all its action immediately. For an analog controller, it should be placed in automatic for a time interval of about 1/20 of the analyzer sample interval; for a digital controller, its scan period should be selected at 1/20 of the analyzer sample interval, and it should be left in automatic for only one scan period. This produces a step output from the controller because all the action is concentrated at the time of the new sample, and no action is taken after that. Integral and derivative settings must be reduced by approximately the ratio of on-time to off-time, i.e., 1/20. This reduces derivative gain by the same factor (as described later in this chapter), thereby eliminating spikes.

The result of this sampled control action is shown for the same process in Fig. 4.6. The controller output consists of clean steps having a similar effect to the filtered output of Fig. 4.5. IAE/Δm is 30.3, 18 percent higher than the unfiltered PID controller. Tuning this controller is quite different from that for continuous PID, however. Its

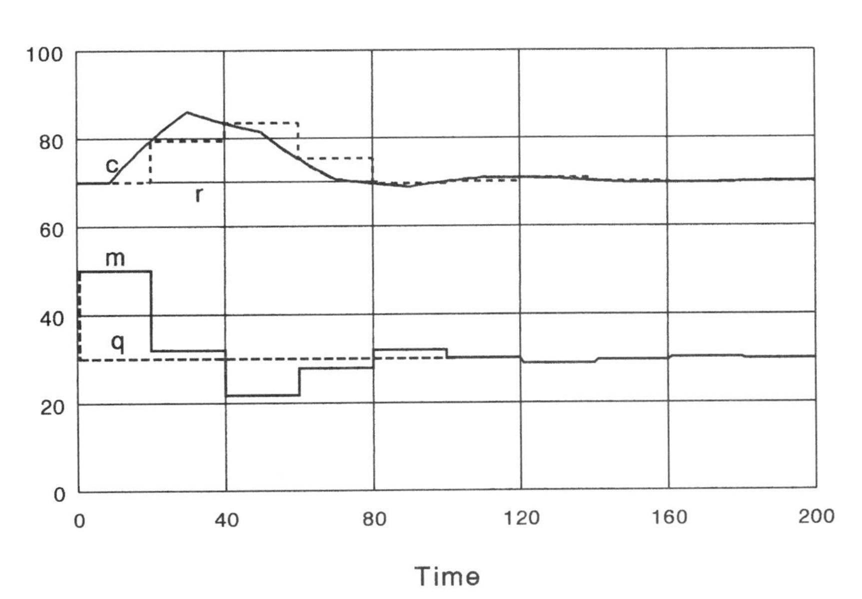

Figure 4.6 Operating the PID controller only when a new sample arrives produces a stepwise output.

proportional band is lower than the others at 80 percent, and I and D are 0.8 and 0.4, respectively. The dynamic settings would be expected to differ from those of the continuous controller by a factor of 20, but in fact, the difference is much greater. Furthermore, the ratio I/D has fallen from 2.4 to 2.0. This is not an easy controller to tune, and there is some inconsistency in the optimal proportional band as a function of the relationship of deadtime to sample interval—it passes through a maximum when they are equal. Tuning rules are developed in detail in Chap. 6.

If the sampling analyzer does not have a synchronizing pulse to tell the controller when a new signal has arrived, a possible substitute is to activate the controller on any change in the controlled variable. Chromatograph outputs are rarely stationary, so this method is workable, but it does have a failure mode. If the measured component is zero or very close to it, the signal may not change from sample to sample, and hence the controller will not take any action. Since a zero measurement always represents a deviation from set point, the controller needs to act but does not. This failure has been observed in distillation column control and will be common there, for the gain of this loop approaches zero as the controlled variable approaches zero. Logic will be required to protect against this eventuality, forcing the controller to follow a fixed time cycle if necessary should the controlled variable remain at zero.

Variations in the sample interval of the analyzer have little effect on the behavior of this loop—in this respect it is quite robust. The settings of integral and derivative time are actually in units of the sample interval. Hence, should the sample interval increase, so do integral and derivative times, as they should. If the analyzer fails, the controller does not wind up, because it remains inactive.

This controller can even be used to control variables measured off-line, such as from the results of laboratory analyses. Each result is entered through the keyboard, and the controller is activated for one calculation as above. More variation may be expected in sample interval when off-line entries are made, which this controller can accommodate, fortunately. Even if a result is lost, the loop remains stable, because no action is taken—the controller is essentially left in manual during the time when no data are available.

Digital PID Controllers

The last controller discussed is effectively a digital controller, even if implemented with analog components. The act of moving the controller output stepwise at regular intervals, whether the controlled variable is measured continuously or not, is the hallmark of a digital controller.

When digital control was first introduced about 1960, proponents expected it to outperform analog control simply owing to more precise calculation capability. As a simple replacement for analog PID control, however, it did not perform as well. The principal reason for poorer performance is the phase lag introduced by sampling. The effect goes deeper than the extension of the period of oscillation, however—derivative gain is seriously compromised as well. And the earliest controllers had no derivative filtering nor windup protection; resolution was not particularly high either, causing either offset or derivative spikes. The best controllers were carefully designed models of their analog counterparts, but they were still limited by their scan period, however.

Both single-loop and distributed digital controllers are in common use. In a single-loop controller, one digital processor has the undivided attention of that loop, sampling at 5 to 10 Hz or even faster. A typical analog process controller has 45 degrees of phase lag at about 10 Hz, but a digital controller sampling at 10 Hz has 180 degrees of phase lag at that frequency without filtering. Add filtering, and the phase lag can easily reach 360 degrees.

In a distributed control system, a single digital processor is shared among a number of loops, and each loop may include several function blocks. The processor is limited in the number of blocks it can process per second. Some distributed systems have fixed scan rates of 2 or 3 Hz, which rigidly limits the number of blocks that can be handled by one processor. To improve flexibility, other distributed systems allow individual assignment of scan intervals to loops, selected from a table of allowable values such as 0.1, 0.5, 1, 5, or 30 seconds. Not surprisingly, it is common practice to maximize the number of blocks processed in an effort to maximize the cost-effectiveness of the system. This results, however, in compromising the dynamic response of the control loops. The $f(\Delta t)$ test needs to be applied, especially to the critical loops in a plant, as a guide to selecting the sample interval. As demonstrated below, however, this test is an inadequate indicator of performance degradation in loops where derivative action is required.

Beyond emulating analog PID controllers, digital technology opens up a whole new world of possibilities: model-based control with deadtime compensation, matrix-based multivariable controllers, fuzzy logic, and self-tuning as well. These issues are taken up in later chapters. The discussion here is limited to implementation and performance differences between analog and digital PID control.

Incremental and positional algorithms

All the controllers described to this point in this text have been positional, in that their output represents the desired position of the manipulated variable. The earliest digital controllers, however, used incremental algorithms, calculating the *change* required in the manipulated variable from its present position. An integrator was used to accumulate these changes, converting them into a position.

The incremental algorithm has several features that distinguish it from its positional counterpart. First, a continuous signal is not required to maintain a valve in a given position—a signal failure then leaves it in its last position. This opens up a discussion on whether fail-safe or fail-in-position is the preferred failure mode for a valve, and there are arguments for both.

If the integrator is at the final actuator, such as a stepping motor, then the operator has no indication of its position, unless additional wires are run to bring that indication back to the control room. However, controller windup is prevented, because there is no integrator in it; the integrating motor drives the valve to its limit and simply ignores any further pulses in the limited direction while responding to pulses in the other direction.

While this is effective for a single loop, it poses a difficult problem in a multiloop configuration. In a cascade system, for example, the incremental output from the primary controller has to be integrated to produce the set point for the secondary controller. If the secondary output should reach a limit, it will not windup. The integrator between the two controllers can, however, and it has to be stopped by logic responding to the limited position of the final actuator.

Another multiloop configuration has two or more controller outputs vying for the right to manipulate a single variable based on the greatest need. The preferred method for controller selection is based on the lowest or highest output (position). If incremental outputs are compared rather than positional outputs, the result is not the same. Since all increments are normally zero, noise on one output tends to be rectified by the selector, incrementing the actuator in the selected direction until another controller produces equal increments in the other direction. The result is that neither controller is able to maintain set point, both controlling with an offset proportional to the noise level.

In the incremental form, the noninteracting PID controller could be implemented as follows:

$$\Delta m = \pm\frac{100}{P}\left[(e_n - e_{n-1} + \frac{\Delta t}{I} e_n - \frac{D}{\Delta t}(c_n - 2c_{n-1} + c_{n-2})\right] \tag{4.11}$$

where subscript n indicates the present value of the variable, $n - 1$ is its last value, etc. The proper sign must be selected to provide the controller action required for negative feedback. Recognize $\Delta t/I$ as a forward-difference operator, so I cannot be set to zero, although D can. The derivative term is the difference between the last two changes in c. This algorithm has no derivative filtering, its derivative gain plainly being equal to $D/\Delta t$. To eliminate proportional response to the set point, $e_n - e_{n-1}$ would be replaced with $c_{n-1} - c_n$.

Proportional and PD control are *not* achievable with this algorithm, since the integrator is inherent in the loop. Even if the integral mode is turned off, there is no possibility of introducing a fixed bias, which is an essential element of these controllers.

The positional noninteracting PID algorithm is simply the integral of the preceding equation:

$$m = \pm\frac{100}{P}\left[e_n + \frac{\Delta t}{I}\sum_{0}^{n} e - \frac{D}{\Delta t}(c_n - c_{n-1})\right] + C \tag{4.12}$$

where C represents the constant of integration. This algorithm *is* convertible to proportional or PD control by turning off the integral mode and replacing C with a fixed bias b. Again, no derivative filtering is shown.

A complete interacting PID algorithm is given next, including derivative filtering and windup protection. Its implementation requires the solution of a succession of equations treating each of the control modes in turn, plus output limits. First, the derivative filter is implemented:

$$c_f = c_f + (c - c_f)\frac{\Delta t}{\Delta t + \alpha D} \tag{4.13}$$

where c_f is the value of c filtered by time constant αD using the backward-difference operator. The derivative term is simply the difference between the input and output of the filter:

$$\Delta c = c - c_f \tag{4.14}$$

Next, the proportional and derivative terms are combined to produce the positional output, using bias term b:

$$m = b \pm \frac{100}{P}(r - c - \Delta c/\alpha) \tag{4.15}$$

where r is the controller's set point. Again, the sign is selected for negative feedback. The calculated output is then compared with its high limit m_h and low limit m_l in a conditional statement:

IF $m > m_h$ THEN $m = m_h$ ELSE IF $m < m_l$ THEN $m = m_l$

Finally, integration is achieved by feedback of the output through a lag representing the integral time, as in the block diagram of Fig. 3.6.

This produces the bias term used in the last equation:

$$b = b + (f - b)\frac{\Delta t}{\Delta t + I} \tag{4.16}$$

where f is normally the limited output m but could come from another source as required to prevent windup. As mentioned earlier, the use of the backward-difference operator produces an effective integral time of $I + \Delta t$, which must be taken into account when estimating integrated error from the controller settings.

The windup-protection feature is extremely valuable, and the noninteracting algorithm lacks it. It is actually easier to convert the preceding interacting controller to noninteracting than to add windup protection to the noninteracting controller described in Eq. 4.12. This is done by augmenting the controller output m with the derivative term before applying it as integral feedback:

$$f = m \pm \Delta c\frac{100}{\alpha P} \tag{4.17}$$

with the sign selection matching that of the Eq. 4.15. As before, feedback term f can be switched to another source for windup protection, but when Eq. 4.17 is operating, the controller is noninteracting.

Derivative gain limit

There are two limiting factors on derivative gain in the noninteracting controller. One is the filter, whose purpose is to limit the dynamic gain to $1/\alpha$. The other is the term $D/\Delta t$ appearing in Eq. 4.12. The limit is the combination of these two factors:

$$K_D = \frac{1}{\alpha + \Delta t/D} \tag{4.18}$$

If the purpose of the filter is to provide a constant value of K_D, then α can be selected to accomplish this (within limits) as a function of D and Δt. The following logic attempts to keep K_D at a value of 10:

$$\text{IF } dt/D < 0.08 \text{ THEN } \alpha = 0.1 - dt/D \text{ ELSE } \alpha = 0.02$$

This method is effective to a lower limit of 0.02 for α, below which K_D falls below 10, following Eq. 4.18.

For the *interacting* controller, a further reduction in K_D accrues as a function of D/I, according to Eq. 3.30. The combination of all three limiting factors gives

$$K_D = \frac{1 + 1/(\alpha + \Delta t/D)}{1 + D/I} \tag{4.19}$$

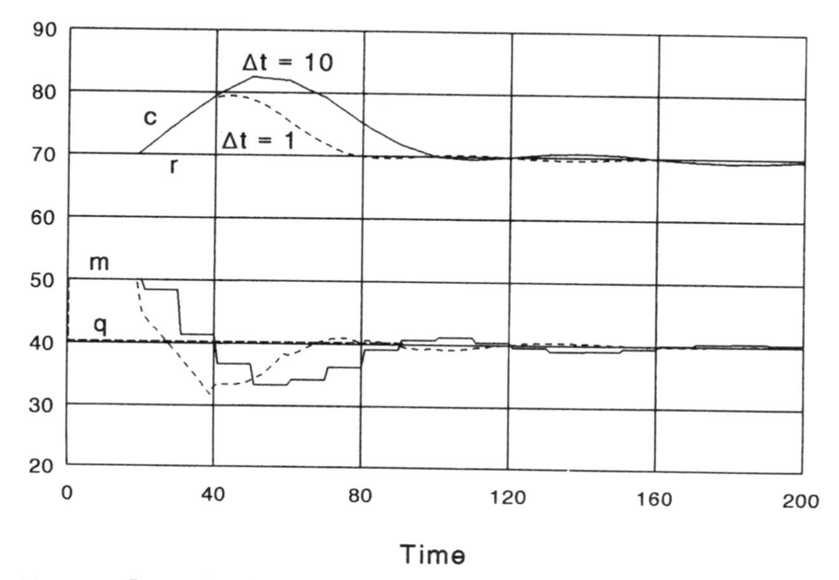

Figure 4.7 Increasing the sample interval of a PID controller sacrifices derivative gain.

When K_D falls appreciably below 10, the performance of the controller suffers, and some readjustment of the PID settings is required to maintain minimum IAE. In Fig. 4.2, increasing the sample interval for a PI loop from $\tau_d/20$ to $\tau_d/2$ was seen to increase IE by a factor of 57 percent. In Fig. 4.7, the same process is controlled using interacting PID controllers with those same sample intervals, both tuned for minimum IAE. IE/Δm for the fast PID controller is 30.2, compared with 61.0 for the fast PI controller. But the integrated error for the slow PID controller is 55.0, an increase of 82 percent caused by the slower sampling, and only 11 percent lower than the fast PI controller.

In addition to the $f(\Delta t)$ factor affecting the loop period and process gain, sampling also has lowered the derivative gain limit. Comparing the settings for the two PID controllers will reveal their derivative gain limits. The fast controller has PID settings of 112, $1.3\tau_d$, and $0.56\tau_d$, respectively, while the slow controller has settings of 145, $1.4\tau_d$, and $0.7\tau_d$; α is limited at 0.02 for both. Using Eq. 4.19, K_D is found to be 7.09 for the fast controller and 1.57 for the slow controller.

Sampled integral controller

The sampled integral controller or digital integrator has some interesting characteristics and useful applications. Its algorithm is very simply:

$$m = m \pm (r - c)\,\Delta t/I \tag{4.20}$$

Here, the forward-difference operator is used, since no feedback loop is formed that can go unstable, as there is in a filter. If I is set equal to Δt, the output moves an amount equal to the deviation during each sample interval. With this setting and the + sign selected, the device can be used as a repeater by connecting an input signal to its set point and its output to the controlled-variable input.

The device also can be used as a very effective controller for deadtime-dominant processes. If its sample interval is long enough to allow the process to reach a steady state following its last action, then deadbeat response is possible. A step-load change will produce a deviation that is acted on by the controller to produce a matching step in the manipulated variable. Following the passage of the process deadtime and before the next sample, the deviation returns to zero. The controller next sees a deviation of zero and takes no further action—deadbeat response.

This controller has two settings that must be matched to the process: Δt must be greater than process deadtime τ_d, and integrator gain $\Delta t/I$ must be the inverse of process gain K_p, i.e., $I = K_p \Delta t$. If the integral time is too short but the sample interval is long enough, the controller output will overshoot the load, developing a cycle having a period of $2\Delta t$ and decay ratio that is the loop gain less 1.0. In other words, the controller gain would have to be twice the optimum to produce an undamped cycle. Too low a controller gain would produce an exponential return to set point. Therefore this loop is quite robust with respect to loop-gain variations.

If the sample interval is too short while the gain is correct, however, the controller does not wait for the deadtime to elapse but makes a second correction before the first response appears. The double correction raises the loop gain to 2, producing an undamped cycle having a period of $4\Delta t$. Thus there is no robustness at all for the sample interval decreasing below the deadtime on a pure-deadtime process. However, since the loop is robust if the sample interval is longer than the deadtime, it should be set for the longest expected deadtime.

Uncertainty in load responses

The principal limitation to sampled integral control is the lack of synchronization between load disturbances and the controller cycle. There is no problem with set-point changes in this regard, for the controller can be made to respond to them immediately. However, the controller is unaware of load disturbances until they affect the controlled variable. A deviation could develop from a load change just before the controller acts, just after, or at any other time during the sample interval. The most favorable appearance of the deviation is just before the controller acts so that it may act as quickly as possible. Conversely, if the deviation develops just after the controller has last acted, almost an entire additional sample interval will elapse before any action is taken. In other words, the IE following a load change to a pure-deadtime process could be as low as the best possible for the most favorable timing and as high as twice the best possible for the least favorable timing.

In Fig. 4.8, three different controllers are applied to a pure-deadtime process. The load response for a continuous integrator is shown, tuned to minimum IAE; its integral time is $1.6K_p\tau_d$, which is also its IE/Δm. Superimposed is the response of a PI controller also tuned to minimum IAE; its IE/Δm is $1.25K_p\tau_d$. The dashed overlaid responses were produced by a sampled integral controller whose sample interval and gain matched the process precisely for the two cases of most favorable and least favorable timing of the disturbance. Its IE/Δm for the most favorable (and best possible) is $1.0K_p\tau_d$, and for the least favorable it is twice that value; on average, it could be expected to fall at $1.5K_p\tau_d$, little better than the integral controller. Furthermore, if the sample interval is lengthened to exceed τ_d for robustness, then IE/Δm could range from $K_p\tau_d$ to $K_p(\tau_d + \Delta t)$, averaging halfway between these wider limits.

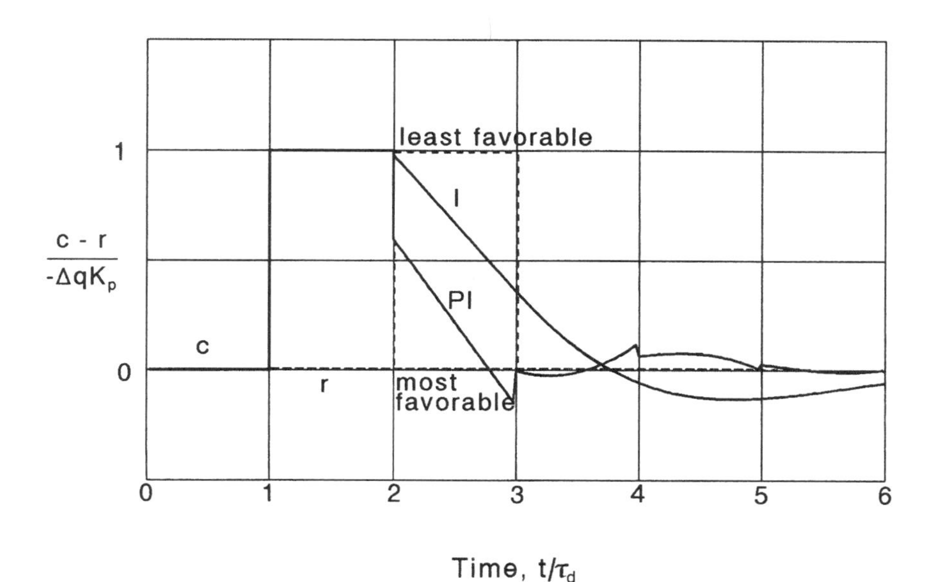

Figure 4.8 Setting the sample interval of a sampled integral controller to equal process deadtime can approach best possible response, but only if the load change occurs at a favorable time in the sampling interval.

This uncertainty can be minimized by breaking the fixed cycle of the controller, attempting to synchronize it with the appearance of the deviation. Leave the controller in manual—"asleep"—until a significant deviation appears. Then let it "awake" and take action, at the same time starting its clock. At the end of the sample interval, it is free to act again, as long as a significant deviation remains. When the deviation is no longer significant, the controller again falls "asleep." In this way, its clock is reset by the appearance of a deviation, and it is effectively synchronized to the load change.

Uncertainty plays a role in all sampled loops, although the pure-deadtime process is most sensitive to it. With increasing lag dominance, especially as a function of the ratio of the primary time constant to the sample interval, it contributes a smaller fraction to the integrated error of the loop. While an optimally tuned digital integrator on a deadtime process produced as much as 100 percent variation in IE, tests using an optimally tuned PID controller on a non-self-regulating process with deadtime produced only a 1.5 percent variation.

Resolution and Threshold Problems

Sampling is not the only impediment to satisfactory PID control using digital technology. Among the first problems encountered was limited resolution in 8- and 12-bit analog-digital converters and processors. However, attempts to control traffic on data buses by going to change-driven communication introduced new problems. These are discussed below in reference to the symptoms they developed.

Limit cycling

If a manipulated variable is unable to match the load perfectly in a non-self-regulating process, the mismatch will cause the controlled variable to ramp up or down, depending on the direction of the mismatch. A controller acting on this ramp will attempt to reposition the manipulated variable to return the deviation to zero. If, due to limited resolution, the manipulated variable must be either higher or lower than the load, however, a limit cycle will result. (This is the same symptom observed when the load happens to be below the lower controllable limit of a valve.) The amplitude of the cycle is determined by the resolution limit and the gain of the process at the period of the cycle. The cycle tends to be sawtooth in appearance, symmetrical if the load happens to be centered between the two nearest positions of the manipulated variable and unsymmetrical and also slower if not.

If the process is self-regulating, the mismatch will produce offset. Without integral action in the controller, the offset will persist, but in attempting to eliminate the offset, integral action will produce a limit cycle. In other words, an integrator anywhere in the loop will produce a limit cycle in the presence of limited resolution. This is not a significant issue when positioning actuators with 12-bit resolution (1 part in 4096, or 0.025 percent), but it definitely is with 8-bit resolution (1 part in 256, or 0.4 percent).

Change-driven communication is sometimes used to minimize traffic on a data bus. However, it imposes a limit that is much more coarse than the inherent resolution of the signal. For example, a threshold as high as 1 percent may be imposed on the controlled variable—any change less than this is not sent to the controller. This naturally results in a limit cycle having an amplitude approximately equal to the threshold. The only remedy is to reduce the threshold to the point where the limit cycle is of an acceptable amplitude. Of course, this defeats the purpose of change-driven communication by increasing the load on the bus to where it would be without this feature. Unfortunately, any attempt to open a loop containing an integrator results in windup followed by overshoot when the loop is again closed. This is the recipe for a limit cycle.

Offset

Limited resolution in the integral term of the controller can stop integration even when a significant deviation persists. The integral gain appeared in Eq. 4.12 as $\Delta t/I$. If the product of integral gain and deviation is less than 1 bit, it may be rounded to zero and therefore fail to add to the integrated output. Consider a controller having a sample interval of 1 second with an integral time of 1 hour—integral gain is 1/3600. In a 16-bit processor, resolution is typically 2^{15} (1 bit is reserved for the sign), which is 1 part in 32,768. Any deviation less than 11 percent will produce less than 1 bit when divided by 3600, which is a much larger offset than one would expect with even a poorly designed analog controller.

There are several possible remedies for the offset problem, which different controller manufacturers have applied. The obvious one is to use double-precision arithmetic for the integration term, but this requires extra memory and time. Another approach is to dither the single-precision result with a random number added in double precision—the result is the same as double-precision integration.[1]

Another approach used by at least one manufacturer is to increase the sample interval automatically as integral time is increased, keeping integral gain constant, at least over most of the range. For example, an integral time of 1 hour would be accompanied by an unrequested sample interval of 30 seconds. While accomplishing the effect of reducing the offset estimated above by a factor of 30, it also changes the loop dynamics. One of the tests commonly performed in gathering information to tune a PID loop is to force a proportional cycle. This is done by setting integral time to maximum and derivative time to zero and reducing the proportional band until an undamped cycle develops. The period of the cycle and the proportional band value are used in the tuning rules. Applying this method to the controller in question will result in a maximum sample interval during testing—far different from that used when the controller is integrating normally—producing misleading test results.

Derivative spiking

Derivative action applied to a step produces a pulse or spike, as shown in Fig. 4.4. In that example, the steps were large because the sample interval was long compared with the process time constant, and the controller was operating on a much faster cycle than the sampling analyzer. One solution to that problem was applied in Fig. 4.6. Derivative gain was reduced to 0.4 by matching the sample interval of the controller to that of the slower analyzer.

If the steps can be kept small, however, the spike amplitude may be acceptable even with the full derivative gain of 10. And this is actually the case with most digital control loops. For example, most single-loop digital controllers use a 12-bit analog-digital converter giving a resolution of 1 part in 4096, or 0.025 percent. These small steps, when multiplied by a proportional gain of even 4 and a derivative gain of 10, will produce spikes of only 1 percent on the controller output. This is probably an acceptable level and should not overly exercise most actuators.

However, I applied two of these controllers configured in cascade to control temperatures in a batch reactor. When properly tuned, each controller had a proportional gain of 5—quite common for temperature. The smallest change in the primary temperature—0.025 percent—was multiplied by $5 \times 5 \times 10$, or 250, producing a train of derivative spikes of 6.25 percent amplitude, even in the steady state. A nonlinear filter was applied (as described in Chap. 8) to suppress the spikes for small deviations, but they always reappeared during upset conditions. Dynamic filtering was avoided as introducing too much lag into the loop.

One technique for minimizing the resolution problem on controlled variables is to use an integrating analog-digital converter. The analog signal is first converted into a proportional frequency which drives a counter. At the end of the sample interval, the total count is used as the digital input but is scaled proportional to the sample interval; then the counter is reset. If the sample interval is short, as on a flow measurement, resolution is lower than it would be on a temperature measurement having a longer sample interval. However, the flow controller would have a low gain and no derivative action, so the low resolution

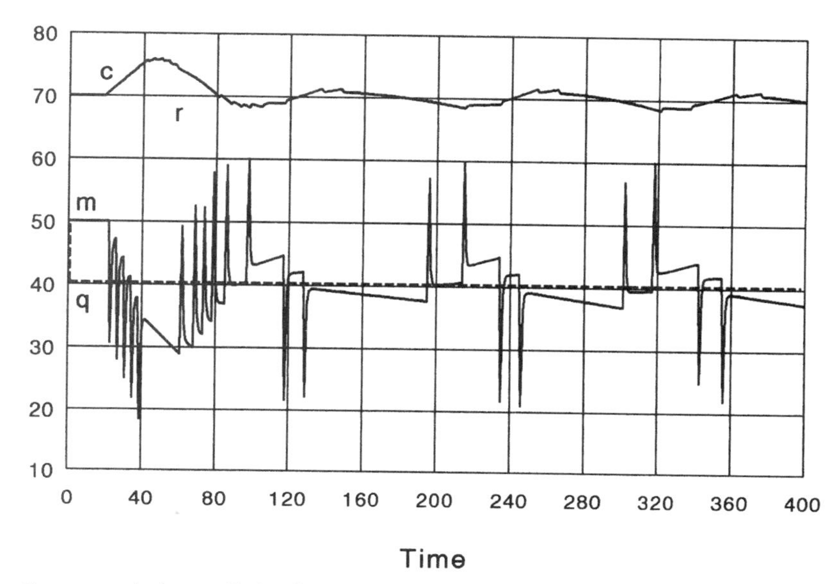

Figure 4.9 A change limit of 1 percent applied to communication of the controlled variable causes both derivative spikes and a limit cycle.

poses no problem there. Meanwhile, the temperature controller with its higher gain and derivative action would be operating on a signal with much higher resolution. The only drawback to this technique is that the reported variable is averaged over the sample interval, which introduces a delay of one-half sample interval. This, of course, doubles the phase lag of the sampler to that of an equal deadtime.

Change-driven communication causes as much grief here as it did in producing a limit cycle with the integral mode. The steps in the controlled variable are the size of the threshold—much larger than the resolution of the signal—thereby amplifying the derivative spikes. Figure 4.9 shows the simulation results for a non-self-regulating process responding to a load change under PID control, with a change limit of 1 percent imposed on the controlled variable. The spikes are that 1 percent multiplied by the product of proportional and derivative gains; the proportional band is 54 percent.

Notation

b	Output bias
c	Controlled variable
c_f	Filtered value of c
c_n	Present value of c

c_{n-1}	Value of c at last sample
C	Constant of integration
d	Derivative operator
D	Derivative time
e	Deviation, 2.713
e_n	Present deviation
e_{n-1}	Deviation at last sample
f	Function, feedback signal
I	Integral time
IAE	Integrated absolute error
IE	Integrated error
K_D	Derivative gain limit
K_p	Process steady-state gain
m	Manipulated variable
m_h	High limit of m
m_l	Low limit of m
P	Proportional band
q	Load
r	Set point
t	Time
Δt	Sample interval
w	Intermediate variable
x	Filter input
y	Filter output
α	Ratio of derivative filter to derivative time
Δ	Difference
τ	Time constant
τ_d	Deadtime
τ_f	Filter time
τ_o	Period of oscillation
ϕ_Δ	Phase shift due to sampling

References

1. Bristol, E. H., "Designing and Programming Control Algorithms for DDC Systems," *Control Eng.*, January 1977, pp. 24–26.

Chapter

5

Model-Based Controllers

The first model-based controller was probably the now-famous *Smith Predictor,*[1] introduced around 1959. It consisted of a first-order model with deadtime in parallel with the modeled process; the controller used was a conventional PID. The Smith predictor at first had very little impact on the world of process control, perhaps because the modeling of deadtime was extremely difficult with the components of the day. Electronic PID controllers were in their infancy, and even obtaining integral time constants in the range of minutes was no small accomplishment. In any case, neither Smith nor his predictor are mentioned in the *Instrument Engineer's Handbook* of 1970[2] or 1985.[3]

The first applications of digital control in the early 1960s were principally PID replacements, batch scheduling, and feedforward control. Innovations in digital model-based control may have begun with Dahlin[4] and his work on paper-machine control later in the decade. Multivariable model-predictive control work began at almost the same time on two fronts: Richalet's *IDCOM*[5] was being applied to refinery operations in France, while Cutler and Ramaker were developing *dynamic matrix control*[6] for Shell Oil Company in the United States. Meanwhile, university research continued in the areas of internal model control and model-predictive control. These techniques are now considered to be at the forefront of process control, overshadowing more mature methods such as PID and feedforward control. They are described in their single-loop embodiment in this chapter to compare performance and robustness against PID control and against best practical control.

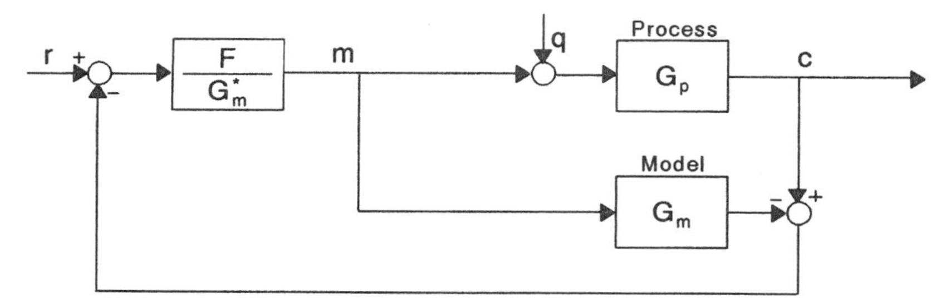

Figure 5.1 In internal model control (IMC), the "controller" consists of an inverse process model without deadtime but with filtering.

Internal Model Control (IMC)

The concept of internal model control has been popular in academic circles for some time. Since it requires inverting the process model, however, practitioners such as myself have not held out much hope for its useful application. Deadtime is not invertible at all, but there are also problems inverting a lag—its inverse, a pure lead, cannot be implemented without an accompanying lag. This has already been demonstrated in the case of derivative action, which as a lead requires a filter or similar method of limiting its dynamic gain.

Figure 5.1 shows the arrangement of blocks required for internal model control (IMC).[7] Manipulated variable m passes through process G_p and a model of the process G_m in parallel. The output of the process is the controlled variable c, whereas the output of the model is an estimate of the controlled variable $\hat{c}$. The accuracy of the estimate depends not only on the faithfulness of the model but also on the load, in that the load enters the process but not the model. (If the load were to enter the model, this would constitute feedforward control, which may be quite viable but is not an essential component of IMC.) If the model is quite accurate, the difference $c - \hat{c}$ will be indicative of the current influence of the load on the controlled variable. Because the two outputs are subtracted prior to feedback, there are two feedback loops—one negative through the process and the other positive through the model.

Pole cancellation

Internal model control can be quite confusing to those familiar with PID control for two reasons. First, the set point r at the input of the "controller" is *not* compared with the controlled variable c, but with the model error $c - \hat{c}$. The result of this comparison is therefore not the deviation e, which must be driven to zero, but the deviation plus the model output, which would only approach zero with zero load.

Second, the "controller" is not familiar but is in fact a lead-lag compensator. Ideally, it is the inverse of the process. In root-locus terminology, each lag in the process is identified as a *pole,* which is to be canceled by a *zero,* or inverse pole in the controller. Consider the simplest practical case of a first-order process with deadtime:

$$G_p(s) = \frac{c(s)}{m(s)} = \frac{K_p e^{-\tau_d s}}{1 + \tau_1 s} \tag{5.1}$$

where K_p is the process steady-state gain, τ_d is its deadtime, and τ_1 is its time constant, s being the LaPlace operator. The process model would contain these same elements, but its inverse could not, since deadtime is not invertible (requiring a time advance):

$$\frac{1}{G_m^*(s)} = \frac{1 + \tau_1 s}{K_p} \tag{5.2}$$

where the asterisk indicates that deadtime has been omitted from the model G_m.

Furthermore, the resulting lead must be accompanied by a lag, which takes the role of a filter:

$$F(s) = \frac{1}{1 + \tau_f s} \tag{5.3}$$

The filter and inverted model are shown combined in Fig. 5.1 to make up the controller. The filter time constant τ_f is intended to be the only tunable parameter in the system, since it does not have a corresponding process component. All other parameters are to be matched as closely as possible to the process. If the process were to be multiorder, the inversion would be more difficult to achieve, although still possible if accompanied by a filter of the same order. The problem remains, of course, to identify accurately a multiorder process. As demonstrated in the following chapters on tuning, identifying lags beyond second order is quite difficult.

Unfortunately, the one pole that cannot be canceled by the IMC is a pole at the origin—an integrator. A process integrator has two inputs: the manipulated variable and the load. In Fig. 5.1, the process model has no load input. Without the load input, an integrator in the model would soon saturate and be of no further use. Even if it had a measured load input, it would eventually saturate, an inevitable result of integration. As a consequence, IMC cannot accommodate a process integrator—a serious limitation, since process integrators (non-self-regulating processes) are quite common in industry. In fact, this is a problem with *all* model-based control technology.[8] An integrating process can still be controlled by model-based controllers, as demonstrated later for a Dahlin controller, a Smith predictor, and the PIDτ_d controller, but the process and model are not matched in this case. In other words, this success is not accomplished by using an integrator in the model. Pole cancellation definitely has its limitations, more of which will become evident in load responses and in tuning methods.

Set-point response

The IMC system was intended primarily to produce near-perfect set-point response. Consider for a moment what might be possible if the controller could be a true process inverse. A set-point change would pass through the inverse and the process in series, with the result that it would be perfectly reproduced in the controlled variable. Without the process model in parallel with the process, however, closure of the feedback loop with its gain of unity would produce uniform oscillations. However, the controller output change is applied to both process and model, with their outputs subtracted. If the two match perfectly, there will then be no feedback and hence no oscillation. Perfect set-point response is therefore possible in this configuration.

However, there are some limitations. First, the manipulated variable will saturate in the presence of a pure lead in the controller. Second, the inverse can have no deadtime, leaving process deadtime in the set-point path. The combination of these two unavoidable limitations reduces the set-point response from perfect to best possible, as defined in Chap. 2. Finally, the filter must be implemented for a realizable controller. The ratio of lead to lag in the controller now limits its dynamic gain in response to set-point changes. Saturation of m may not take place if the set-point change is small. In any case, the set-point response will differ from best possible on the basis of the time constant selected for the filter. Without saturation, it will be

$$\frac{c(s)}{r(s)} = \frac{e^{-\tau_d s}}{1 + \tau_f s} \tag{5.4}$$

where filter time constant τ_f is said to be the *closed-loop time constant.* A typical minimum value for τ_f would be $\tau_1/10$, for the same reason that this factor was used for the derivative filter in a PID controller. Furthermore, if saturation does develop, the set-point response will be slower than what Eq. 5.4 would predict.

Load response

The response of the system to load changes depends very much on where the load is introduced into the loop or, to put it another way, on the transfer function in the load path. In most representations of IMC, the load is shown introduced at the process output without any dynamics in the path, simply adding to the controlled variable. If this were true, then the signal fed back to the controller would be the load q. The IMC response to the load change would then be identical to the set-point response described above, approaching best possible, depending on the setting of the filter time constant.

True load changes never enter the loop at that point unmodified by some dynamic elements. In structuring a system, the control engineer is advised to select for manipulation that variable having the most influence on the controlled variable both in the steady and unsteady states. If a load variable were to affect the controlled variable with no discernible dynamics, it ought to be used as a manipulated variable, which would render the regulation problem trivial.

In fact, the load always enters upstream of the dominant process time constant and, in many cases, at the same point as the manipulated variable. Several common applications are here mentioned to make this point:

1. Liquid level (inventory) is the integral of the difference between stream flow rates entering and leaving a vessel—one of these is manipulated, with the rest being loads.
2. Product composition from a distillation column follows the same dominant lag for all disturbances: feed rate, reflux, heat input, etc.
3. Heat exchangers and boilers may have different combinations of deadtime and lag on the two sides of the heat-transfer surface, but they still tend to be similar due to common heat capacity.

Take the common case where the dominant lag is the same for both load and manipulated input, which is shown in Fig. 5.1. Following a step change in load, the controlled variable will depart from set point along an exponential trajectory determined by that dominant lag. Since there is no response from the model, this change in c is fed back to the controller, where it passes through the inverse process model. In the absence of any filtering, the controller would produce a step change in m exactly equal in size to the load step but delayed by the process deadtime.

This output trajectory will produce the best possible load response *only* if the process has no lags at all, as shown in Fig. 2.2. The output motions required for best possible load response for processes having lags—self-regulating and not—all *overshoot* the load, as shown in Figs. 2.3 through 2.6. Therefore, IMC only approaches best possible load response for pure deadtime processes or, more precisely, for those processes having no lag in the load path. In practice, the step

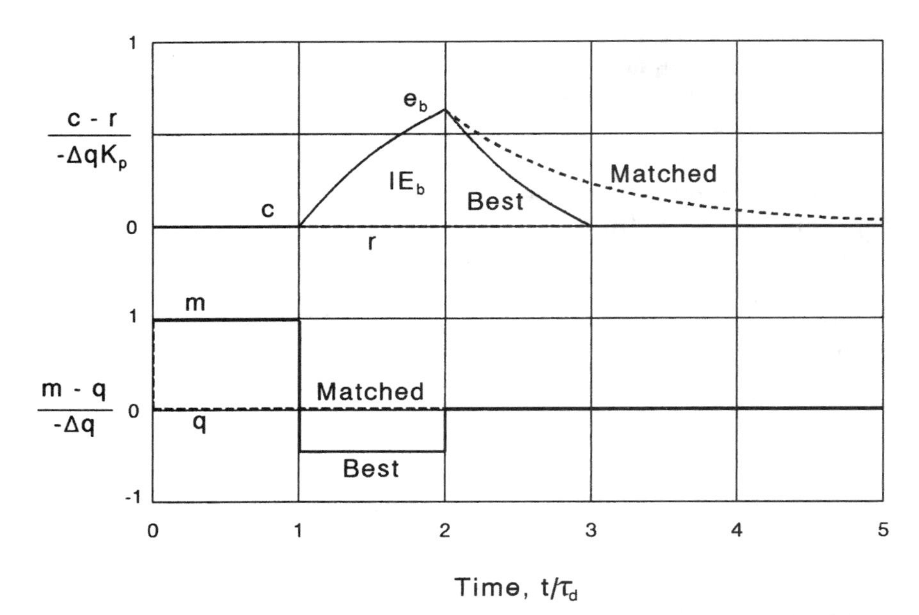

Figure 5.2 Matching the manipulated variable to the load produces an exponential recovery at the time constant in the load path.

change in output is only obtained in the absence of the filter, which interferes further with the response.

For the first-order process with deadtime described above, matching the manipulated variable with the load produced the exponential recovery curve, which is compared with the best practical in Fig. 5.2. The area between the load and the matched manipulated variable is $-\Delta q\ \tau_d$; when passed through the process, the area under the controlled-variable response curve is the same value amplified by process gain K_p:

$$\mathrm{IE}_{\mathrm{IMC}} = -\Delta q\, K_p \tau_d \tag{5.5}$$

which is independent of the lags in the process. When this $\mathrm{IE}_{\mathrm{IMC}}$ is divided into the best practical, we have the practical performance rating of the matched IMC controller:

$$\frac{\mathrm{IE}_{\mathrm{IMC}}}{\mathrm{IE}_b} = \frac{1}{1 - e^{-\tau_d/\tau_1}} \tag{5.6}$$

As is evident, its performance slips badly as τ_d/τ_1 becomes small, i.e., as the process becomes lag-dominant.

Adding the necessary filtering only makes performance worse, for it increases IE in direct proportion to the filter time constant:

$$\mathrm{IE}_{\mathrm{IMC}} = -\Delta q K_p(\tau_d + \tau_f) \tag{5.7}$$

Comparisons with PID controllers are saved for later in the chapter, after other model-based controllers are described. Unfortunately, there is no easy remedy for the exponential recovery, since there is no easy way to produce the overshoot required for better load response without compromising set-point response. Other model-based controllers have somewhat more flexibility in improving load response owing to their structure. The actual structure of IMC must be modified to improve its load response.

However, it is instructive at this point to examine the transfer function of IMC to a load change. If the controller and the model that parallels the process are combined into a single function, the response of the controller output m to any load-induced change in the controlled variable c can be evaluated in terms similar to those of a more familiar controller such as a PID. This method is used here to compare the various model-based controllers functionally and to identify the parameters that contribute the functions. For the first-order self-regulating process with deadtime, the IMC transfer function is found to be:

$$\frac{m(s)}{c(s)} = -\frac{1 + \tau_1 s}{K_p(1 + \tau_f s - e^{-\tau_d s})} \tag{5.8}$$

The appearance of the process time constant in the numerator identifies it as a pole-canceling device. The deadtime function in the denominator will be found common to all model-based controllers. And finally, the filter time constant in the denominator is the only tunable term in the function. This "one-knob" tuning is considered desirable in freeing plant engineers from the very demanding and complex task of controller tuning and is considered to be a distinct advantage of model-based control over PID control. Whether it is indeed an advantage or an impediment to performance is an issue to be examined in detail.

In any case, IMC has been introduced and analyzed first because it is the simplest of the model-based controllers functionally, yet it embodies both the theoretical advantages and limitations of the method. The other controllers that follow in this chapter are quite similar in function yet bear different names and are implemented differently. Each, in turn, will be compared with IMC to relate to it functionally but also to see how it differs in implementation and flexibility in hopes to overcome the poor load response described above.

The analytical predictor

In an effort to correct the load-response limitation of IMC, Doss and Moore[9] conceived of a method that would estimate the magnitude of the load change and pass it along to the controller for action. Recall that in Fig. 5.1 the signal traveling the feedback path to the controller is the difference between the true value of the controlled variable and its value estimated by applying the controller output to the process model. If the model faithfully represents the process, then that difference is

$$c - \hat{c} = qG_q \tag{5.9}$$

where q is the load, and G_q is used in place of G_p to include the more general case where the manipulated and load variables may enter the process at different points and therefore display different dynamics. To improve load response without affecting set-point response requires the insertion of a dynamic compensator in this path. This compensator is given the name of an *analytical predictor.*

In the simplest case, the load dynamics could be represented by deadtime and a first-order lag. There is no need to compensate the deadtime, because, as described in Chap. 2, it has no effect on the resulting response curve. The reason load response with IMC is so poor is the *lag* in the load path, whose time constant determines the exponential return to set point. Recall that if a step-load change were introduced at the process output, recovery would be as fast as the set-point response, i.e., limited only by the filter and any saturation. Therefore, the ideal analytical predictor is $1/G_q^*$, where the asterisk indicates the omission of any deadtime in G_q. As with the controller function, however, a lead cannot exist alone, so a practical analytical predictor is in fact a lead-lag compensator with dominant lead.

The digital implementation of the predictor amounts to estimating the load change based on predicting its trajectory N time steps ahead, on the basis of the most recent (unrequested) change in c, where N represents the number of samples making up the deadtime in the manipulated path. Experiments conducted with such a predictor reported load-recovery curves exceeding the best practical and approaching the best possible.[10] In other words, recovery from peak deviation was achieved in less than one deadtime by forcing the manipulated variable to overshoot the load more than the best practical shown in Fig. 5.2, even to the point of saturation.

The question of practicality now arises. If the model G_m were perfectly matched with the process G_p, the feedback loop would remain open, carrying only load signals to the controller in an open-loop manner. In this case, the gain between the load input and the manipulated-variable output would not matter—any desired value could be used. However, any mismatch between the model and process will allow a remnant signal to be passed back around the loop. IMC loops already have a gain of 1.0, as required to provide the desired set-point response, which is enough to start an undamped cycle in the presence of a deadtime mismatch depending on the level of filtering. The added dominant lead required by the predictor raises the dynamic loop gain further so that even a mismatch in gain or lag is likely to produce an expanding cycle. And there is still deadtime in the loop as well, so the pattern will tend to be repeated or even reinforced each deadtime. While the analytical predictor is capable of a dramatic performance improvement on load changes, it is likely to fail robustness tests, either to be bypassed or to be filtered into ineffectiveness.

Direct Synthesis Controllers

Another approach to the same problem is the formulation of a controller that forces a given process to satisfy a specified set-point response. For example, let the process given by Eq. 5.1 be required to respond to set-point changes in the manner described by Eq. 5.4. The set-point response of a closed loop comprised of a controller G_c and a process G_p must follow the general formula

$$\frac{c(s)}{r(s)} = \frac{G_c G_p}{1 + G_c G_p} \tag{5.10}$$

where $G_c(s)$ is the controller transfer function $m(s)/e(s)$. When Eq. 5.10 is set equal to the desired set-point response of Eq. 5.4, a solution may be found for the controller transfer function. Interestingly, $G_c(s)$ comes out to be exactly equal to Eq. 5.8, except for a reversal of sign, in that $e = r - c$. Thus the direct synthesis controller is identical to IMC in function but appears as a single block, having both set point and controlled variable as inputs and acting on their deviation as does a conventional controller. The best known of these controllers are implemented digitally, as described below.

The Dahlin controller

The Dahlin controller is simply a digital implementation of the direct synthesis transfer function:

$$m(s) = e(s) \frac{1 + \tau_1 s}{K_p(1 + \tau_f s - e^{-\tau_d s})} \tag{5.11}$$

(Unfortunately, it becomes necessary to use e to represent two different terms in the same equation: As $e(s)$, it is deviation, and as e^x, it is the Naperian base constant 2.718.) We move the dynamic terms from the denominator:

$$m(s)(1 + \tau_f s - e^{-\tau_d s}) = e(s)(1 + \tau_1 s)/K_p \tag{5.12}$$

Next, the result is transformed into the time domain:

$$m(t) + \tau_f \frac{dm}{dt} - m\,(t - \tau_d) = \left[e(t) + \tau_1 \frac{de}{dt}\right]/K_p \tag{5.13}$$

Converting to a difference equation for digital implementation allows a solution of the new value of m in terms of previous values of m and deviation e:

$$m_n = Bm_{n-1} + (1 - B)m_{n-N-1} + K(e_n - Ae_{n-1}) \tag{5.14}$$

where subscript n represents the new value, $n - 1$ is the last value, and $n - N - 1$ is the value $N + 1$ samples ago. It is imperative that N be an integer equal to or exceeding $\tau_d/\Delta t$, or instability can result; the controller deadtime is thereby $N\Delta t$.

Coefficients A and B will be recognized as complements to the exponential difference operators presented in the last chapter:

$$A = e^{-\Delta t/\tau_1} \qquad B = e^{-\Delta t/\tau_f} \tag{5.15}$$

To produce the set-point response upon which the direct synthesis controller is based, gain K must be set to include both difference operators:

$$K = \frac{1 - B}{K_p(1 - A)} \tag{5.16}$$

When the parameters are set into the Dahlin controller to match the first-order process with deadtime, its response is identical to that of IMC if sampling is fast enough. Sampling at one-tenth the deadtime reveals some "ringing" in the controller output, especially in response to set-point changes, as shown in Fig. 5.3 for a first-order self-regulating process having equal lag and deadtime. Response to a step-load change is observed first, followed by a set-point change. Applying lead to the set-point step produces the large pulse, which repeats every deadtime, while attenuated by the filter. In this simulation, τ_f is set equal to the sample interval. If the sample interval is cut in half, ringing disappears altogether, but if it is doubled, the loop becomes unstable until the filter time is increased to equal Δt.

This controller is more flexible than IMC, in that it can be tuned to improve its load response. Figure 5.4 shows the same process being controlled, but with the sample interval and filter time of the controller cut in half and its gain increased to minimize IAE for load changes. The 50 percent higher gain causes the controlled variable to return to set point one deadtime after the peak deviation instead of following the usual exponential recovery curve; a reduction in IAE of

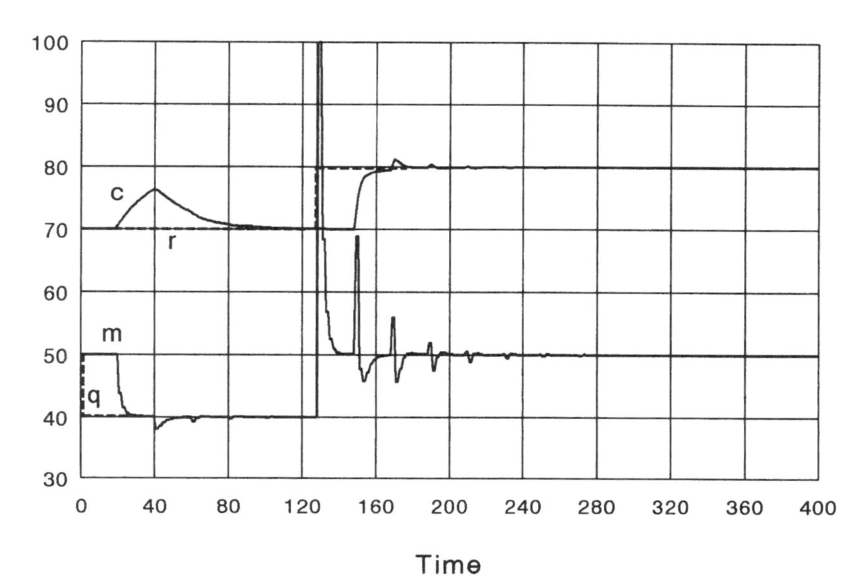

Figure 5.3 The Dahlin controller produces load and set-point responses similar to the IMC controller, but its output "rings" if sampling is slow.

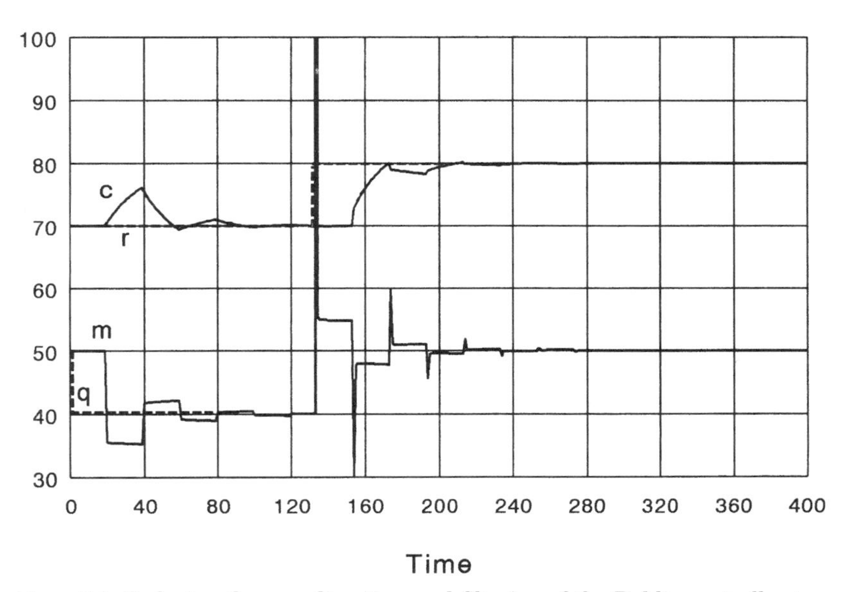

Figure 5.4 Reducing the sampling time and filtering of the Dahlin controller to a minimum while increasing the gain minimizes IAE in response to load changes, but set-point response suffers.

31 percent results, 11 percent above best practical. Unfortunately, the set-point response is unfavorably affected, and the higher gain causes ringing as well.

If one is willing to tune this controller, its use can even be extended to non-self-regulating processes (although a better candidate is the PIDτ_d controller described later). Unfortunately, this version of the Dahlin controller has some interaction between the filter setting and the gain adjustment, as shown in Eq. 5.16. The gain setting depends on the process gain and time constant, as well it should, but it also depends on the filter setting. If the Dahlin controller is reduced back to its transfer function (assuming continuous rather than sampled operation), $1/K_p$ in the transfer function would be replaced with $K\tau_f/\tau_1$:

$$\frac{m(s)}{e(s)} = \frac{K\tau_f(1 + \tau_1 s)}{\tau_1(1 + \tau_f s - e^{-\tau_d s})} \tag{5.17}$$

In other words, whenever the filter is adjusted, the gain requires readjustment as well, an annoyance in that the filter may be considered by some the *only* tuning parameter required. This interaction could be avoided by reformulating the algorithm.

Second-order controllers

A direct synthesis controller is derived from a process transfer function and a realistic expectation for set-point response. If a second-order process is to be controlled, a first-order expectation for set-point response requires second-derivative action. This is not very realistic, and therefore, a first-order controller must fall short of expectations. If the preceding first-order Dahlin controller is applied to a second-order process with deadtime, ringing appears even with fast sampling. Figure 5.5 shows some deterioration in the load response but far more in the set-point response. The first-order controller has four adjustable parameters: A, B, N, and K. To properly satisfy a second-order process according to ref. 11, three more parameters are required to synthesize the second-order digital controller. This level of complexity essentially removes the possibility of tuning. Controller parameters must therefore be set to match process parameters as closely as possible and hope for the best. If the match is less than perfect, or if the process changes from the original matched condition, there is no chance of readjusting the right parameters the required amount to rematch them. Either the process must be identified again, or filtering must be increased for robustness, or both, depending on the degree of mismatch.

Vogel and Edgar have apparently improved on the second-order Dahlin controller with one of their own.[11] However, it is of the same level of complexity, though somewhat more robust, avoiding most of

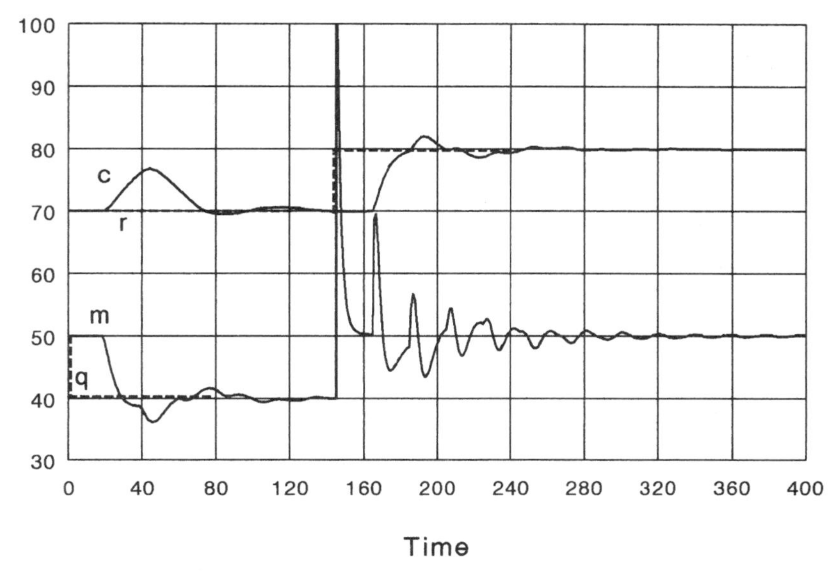

Figure 5.5 The time constant of the process in Fig. 5.4 has been split 3/1 into primary and secondary lags in this simulation, causing the first-order Dahlin controller to ring, even with fast sampling.

the ringing. When second-order terms are eliminated, the Vogel-Edgar controller reduces to the Dahlin. The problem remains to identify the process accurately enough to fill in all the needed parameters.

Leaving the filter and gain adjustments accessible to tuning might allow both these second-order controllers to be tuned for minimum IAE on load changes, as was done for the first-order process in Fig. 5.4. While a first-order Dahlin controller can be tuned to control a first-order non-self-regulating process satisfactorily, it does not perform nor tune nearly as well if that process is second order. Rather than devote any further space to the second-order controllers, the application of model-based control of second-order processes is deferred to the PIDτ_d controller, which both controls them more effectively and tunes more easily.

Model-Predictive Control (MPC)

The first and simplest of the schemes for MPC uses the Smith predictor, a linear first-order model of a process. However, the technology extends well beyond this modest beginning into nonlinear, multiorder, and multivariable plant models. To facilitate comparison of MPC with other single-loop controllers, the treatment that follows will be restricted to linear and single-loop processes but with extension to multiorder processes.

Model-predictive control has properties in common with both IMC and direct synthesis control. It uses a process model in parallel with the process, as does IMC. However, the signal acted on by the controller is a true deviation, at least in the steady state. Thus it can use a conventional controller rather than a lead-lag compensator, which acted as the controller in IMC. Similar to the direct synthesis controller, its gain can be set very high and does reflect the sampling interval as well as the process time constant. However, the implementation of filtering differs from both the preceding controllers. The simplest way to compare them is through the Smith predictor.

The Smith predictor

Figure 5.6 illustrates the block structure of a Smith predictor modified by me to simplify tuning. The essence of the predictor is a process model in parallel with the process, here assuming a first-order self-regulating process with deadtime. However, in contrast with IMC, the predictor has a bypass connection around the deadtime in the model. The model including the deadtime forms a positive-feedback loop that cancels the negative-feedback loop through the process, as done with IMC. But now there remains another negative-feedback loop through the model but bypassing the deadtime. This bypassing signal *predicts* how the process would respond if it had no deadtime, allowing the controller to speed up its corrective action through this minimum-phase closed loop. Lacking deadtime in this remaining loop, the controller gain can be set very high—a function of the controller sampling time rather than the longer process deadtime.

In the steady state, the signals through the model deadtime and its bypass cancel, leaving only the negative-feedback loop through the process. Hence model error has no steady-state component and therefore cannot contribute to offset. In the unsteady state, the predicted response is acted on by the controller, while the two slow loops cancel, assuming no model error. The three blocks in the model could be

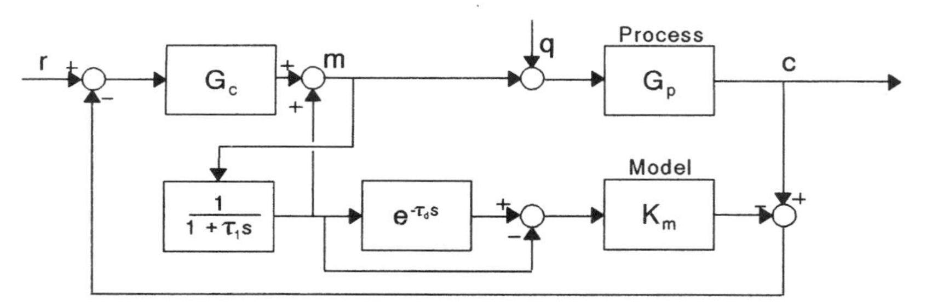

Figure 5.6 This Smith predictor has been modified by positive feedback from the model lag back to the controller output to provide integral action.

arranged in any sequence because they are all linear, but this particular sequence has been selected to facilitate tuning. Because the steady-state input to the model gain block is zero, adjusting K_m on-line will not disturb the loop, and this is the most influential parameter.

For a loop having only a sampler and a lag, the controller gain can be set about at the ratio of time constant to sample interval divided by the process gain. Most proponents of MPC recommend setting the sample interval at about one-tenth the time constant, allowing K_c to be set as high as $10/K_p$. This is still not high enough to avoid the possibility of offset resulting from load changes, however, so integral action is usually required in the controller. Figure 5.6 simply identifies the controller as G_c, which in most applications of the Smith predictor is a PI controller.

However, I have added another positive-feedback loop to the diagram, taken from the output of the process lag and added back to the controller output. This will be recognized as integral feedback. With this modification in place, the model lag also acts as the integral time constant for the controller, leaving G_c with only proportional action. This modification simplifies both the transfer function for the system and tuning of the controller, eliminating one adjustment. The resulting transfer function for response of the manipulated to the controlled variable resulting from load changes is:

$$\frac{m(s)}{c(s)} = -\frac{1 + \tau_1 s}{K_m(1 + \tau_1 s/K_c K_m - e^{-\tau_d s})} \tag{5.18}$$

where K_m has been used to identify the model gain apart from the process gain as a potential tuning adjustment. Observe that the form of the first-order MPC transfer function is identical to that of the IMC function Eq. 5.8. The only difference is that the filter time τ_f in IMC has been replaced with $\tau_1/K_c K_m$. Filtering is still available, through the practice of reducing the controller gain below the maximum allowed value estimated above. The effective filter time becomes Δt multiplied by the ratio of the maximum to the actual controller gain less 1.0.

If the connection added by me is removed and a PI controller is used, the only term in Eq. 5.18 affected is the filter, becoming

$$\tau_f = \frac{Is(1 + \tau_1 s)}{K_c K_m(1 + Is)} \tag{5.19}$$

where I is the integral time of the controller. Since the filter is intended to be a small factor, the additional degree of freedom obtained by independently setting the integral time has little influence on the response of the loop.

As designed, the Smith predictor provides the same set-point and load responses as IMC and a fast-sampling Dahlin controller on a first-order self-regulating process. While the set-point response approaches the best possible (depending on the filter time), recovery from a step-load change entering upstream of the process time constant follows the exponential curve of the time constant. Nor is it intended to be applied to non-self-regulating processes. In other words, as designed, MPC shares all the advantages and disadvantages of the other model-based control methods. IMC was particularly inflexible in this regard, there being no way to improve on the load response without altering the functionality of the system with the analytical predictor or compromising its set-point response.

As evidenced in Fig. 5.5, however, the Dahlin controller could be tuned by increasing its gain to minimize IAE for load changes, albeit set-point response was thereby degraded. The Dahlin controller could even be tuned for good load response on first-order non-self-regulating processes. The Smith predictor can be tuned in a similar way, not by increasing the controller gain K_c, which only affects the filtering, but by *reducing* model gain K_m. This has an effect equivalent to increasing the gain K of the Dahlin controller, which appeared in the numerator of the reconstructed transfer function (Eq. 5.17). In fact, with K_c set to its maximum value to eliminate any filtering, the Smith predictor can be tuned to approach the best practical response for all first-order processes, self-regulating and not.[12] Again, the principal adjustment used to minimize IAE is K_m, although in the application to non-self-regulating processes the model and process are not related, so other adjustments must be made as well. More is said on this subject under the PIDτ_d controller below, whose tuning rules are fully developed in the next chapter.

The curious aspect of MPC is that except for pure-deadtime processes, such poor load response is achieved *as designed.* To improve the load response requires departing from the design by intentionally mismatching the model and process parameters. The case of the non-self-regulating process is extreme, for the model even differs functionally from the process—yet excellent control can be achieved if careful tuning is applied.

Matrix-based controllers

A closer match to the process than is possible with first- or second-order models can be obtained with convolution models. A step or pulse is applied to a manipulated variable, and the resulting response in the controlled variable(s) is marked at regular time intervals until a new steady state is approached. Figure 5.7 identifies 30 coefficients a_1 to a_{30} as the lengths of the lines between the initial value of the

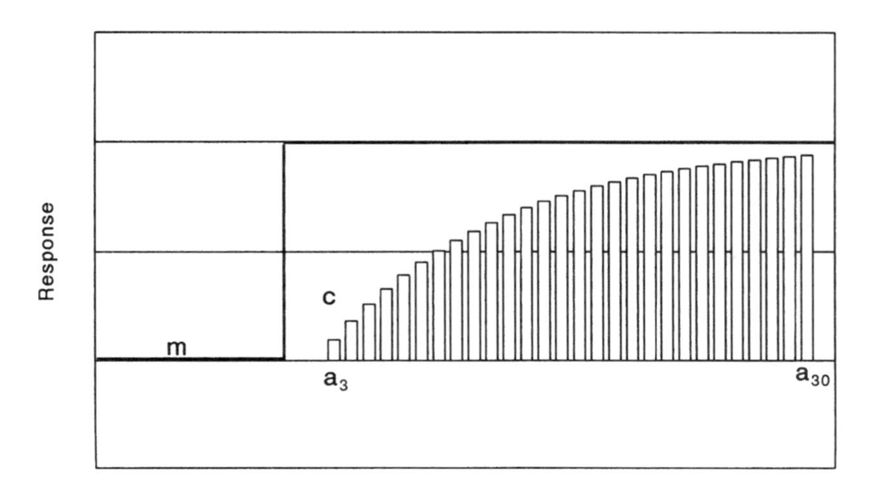

Figure 5.7 A convolution model is a vector representing the response of a controlled variable at regular increments of time between steady states.

controlled variable and its value at each sample interval in response to a step. The example is of a first-order process with deadtime, but the method is applicable to any combination of process dynamics, except an integrator. To model an integrator successfully with this method, the derivative of the controlled variable must be taken.

The pulse response of the same process will reveal no change until the deadtime elapses but then rises to a maximum before decaying exponentially to the original steady state. The coefficients obtained for the pulse response will be the difference between each adjacent pair of step-response coefficients. This allows either model to be used with either method of testing.

If the test is not continued until a final steady state is reached, a discontinuity may appear when the model is run, occurring at time intervals equal to the length of the test. To minimize this problem, the test should cover a 99 percent complete response. The response curve of Fig. 5.7 shows 30 coefficients spread over the deadtime and 2.8 time constants, at the end of which response is only 94 percent complete. This is probably not quite enough for satisfactory performance—4.6 time constants would be required for 99 percent response. However, 30 coefficients seems to be the number commonly used in these models, which would require the sample interval in Fig. 5.7 to be increased from one-half to one deadtime. Unfortunately, this reduces the definition of the deadtime, as well as slowing the response of the control loop.

Having obtained the convolution model, it can be used to predict the response of the controlled variable at any future time up to the *model horizon* (the last coefficient) for any combination of past values of the manipulated and controlled variables. Additionally, a controller based on the same technology can predict a number of moves of the manipulated variable up to the *control horizon* that would be necessary to bring the controlled variable to set point, although only the current move need be actually executed. Prediction of the controlled-variable trajectory including these moves can be made up to some *prediction horizon:*

$$\begin{bmatrix} \hat{c}_1 \\ \hat{c}_2 \\ \hat{c}_3 \\ \vdots \\ \hat{c}_C \end{bmatrix} = \begin{bmatrix} a_1 & 0 & 0 & \dots & 0 \\ a_2 & a_1 & 0 & \dots & 0 \\ a_3 & a_2 & a_1 & \dots & 0 \\ \vdots & \vdots & \vdots & \dots & 0 \\ a_P & a_{P-1} & a_{P-2} & \dots & a_{P-C+1} \end{bmatrix} \begin{bmatrix} \Delta m_0 \\ \Delta m_1 \\ \Delta m_2 \\ \vdots \\ \Delta m_{C-1} \end{bmatrix} \tag{5.20}$$

where subscript C represents the number of sample intervals from the present to the control horizon, and P is the number of samples to the prediction horizon. The triangular matrix of dimension PC is known as the *dynamic matrix.*

The prediction horizon must be far enough to reveal any oscillations that might develop by the inversion of unstable zeros. They are the residue of process deadtime being a noninteger number of samples. In Fig. 5.7, the deadtime happens to be exactly two samples, but this is a contrived case—a noninteger result is most probable, especially with slow sampling. The only sure way to avoid this problem is to select a sample interval longer than the longest anticipated deadtime; unfortunately, this will affect performance severely, as demonstrated later in this chapter.

In practice, the model output is updated each sample interval by comparing it with the present value of the controlled variable and applying the observed difference to all future predictions. If the control horizon is limited to only one move, then the dynamic matrix controller (DMC) operates functionally as a direct synthesis controller. However, by calculating several future moves, it can make larger changes in the next move. As a consequence, its output is capable of large excursions following a set-point change, similar to what the Dahlin controller does. To overcome this tendency, *move suppression* is provided. This is achieved by penalizing the controller for output changes relative to the penalty for deviation. The tuning adjustments for the DMC are the selection of the control horizon and move suppression.

The DMC is capable of outperforming the PID controller on set-point changes but not on load changes introduced upstream of a dominant lag. Morari and Lee[8] propose to correct this limitation by representing the load as a ramp introduced at the process output through a double integration of white noise rather than a single integration, as is done with DMC. Still, on lag-dominant loops, the requirement of selecting a sample interval which is the dominant time constant divided by 5 to 10 means that there will be only one or two samples per deadtime. This in itself can degrade performance of a model-based controller to less than a faster-sampling PID controller, which is demonstrated later in this chapter. In some of the published reports claiming superior performance against PID controllers, mostly set-point upsets are investigated, and the PID controller turns out to be badly mistuned.[13,14]

This technology is most useful in a multivariable environment, resulting in a linear multivariable controller. Because the controller is linear, however, it cannot match a nonlinear plant perfectly and will only perform best at the conditions at which the model was developed. When the plant moves away from these conditions, the model may require updating. While individual nonlinearities such as valve characteristics may be compensated, nonlinear combination of process variables by multiplication or division cannot be compensated for in a linear system.

An important consideration in installing one of these systems is the large commitment of workers and time required to produce the model. Tests must last at least for four time constants for each manipulated variable on a process unit. In the case of a distillation column, there are typically four or five manipulated variables, and time constants of an hour or more are common. Thus testing may require up to a week to develop the model.[15] Morari and Lee[8] also complain that the method is computationally excessive, gives generally sluggish load response, and behaves poorly in a multivariable environment on ill-conditioned systems (those systems having a high level of interaction among loops). This technology is much more effective in a role of feedforward control and decoupling than in feedback control, where it is dynamically inefficient. Philosophically, DMC and IDCOM also violate Albert Einstein's rule that "things should be kept as simple as possible, but no simpler."

The PIDτ_d Controller

Another model-based control method especially applicable to the single-loop environment is the addition of deadtime compensation to a conventional PID controller. There are several advantages to this approach. First, a PIDτ_d controller can be constructed rather simply by adding a deadtime block to an existing single-loop digital controller or to a controller in a distributed system. Second, the controller is tunable, much as a PID controller. Third, a very high level of performance is attainable, superior to MPC and IMC controllers for step-load changes to lag-dominant processes. And last, the control algorithm is not computationally intensive, so sampling can be much faster than possible with DMC, for example. Yet the controller is quite applicable to any process, including non-self-regulating, steady-state unstable, and even those consisting of multiple lags.

Controller structure

Figure 5.8 shows a conventional interacting PID controller in its simplest form, with filtered derivative D applied to the controlled variable and integration achieved by positive feedback of the controller output through a lag representing the integral time constant I. Inserted in this feedback path is a deadtime block. The controller now has effectively four modes and can be operated with any but proportional set to zero. In other words, the controller can act as PI, PID, $P\tau_d$, $PD\tau_d$, $PI\tau_d$, or $PID\tau_d$ simply by setting the unused parameters to zero. When controlling a first-order process with deadtime, for example, I can be set to zero, or for a deadtime process, both I and D can be set to zero. Additionally, in a noisy environment, it may be necessary to set D to zero, leaving a $PI\tau_d$ controller, which is much more effective than a PI controller, giving performance comparable with a PID controller but without the noise sensitivity.

Functionally, the controller compares with the IMC, Smith, and Dahlin controllers, which can be determined by solving the block diagram for the response of output m in terms of controlled variable c:

$$\frac{m(s)}{c(s)} = \mp\frac{100}{P}\frac{(1 + Ds)(1 + Is)}{(1 + \alpha Ds)(1 + Is - e^{-\tau_d s})} \tag{5.21}$$

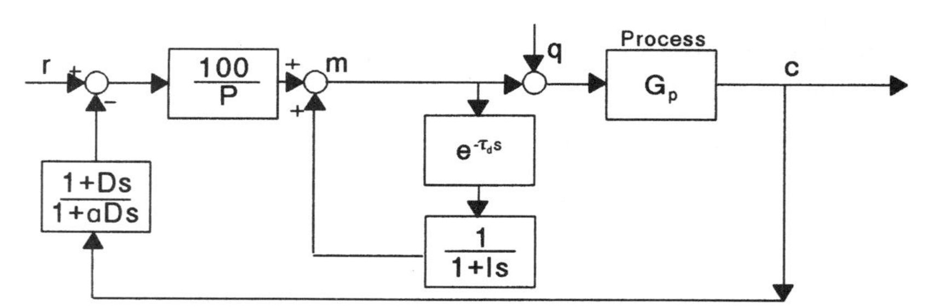

Figure 5.8 The insertion of a deadtime block into the integral-feedback loop has converted this PID controller into a model-based $PID\tau_d$ controller.

Note that this transfer function is left in terms of the controller settings instead of process parameters, as was the case for the IMC and Smith controllers. Their functions had a single lead identified as the process time constant, where this controller has two: I and D. To match the other transfer functions, it is only necessary to set I equal to the derivative filter αD. Then D matches the process lag τ_1, the proportional gain $100/P$ is the inverse of the process gain K_p, and integral time I acts as the filter. The amount of filtering actually applied is determined by α, which is typically 0.1.

This matching of the controller parameters against the process is only illustrative, however, for the PIDτ_d controller is not constrained to operate in this manner, as is the IMC controller. For a first-order lag-dominant or non-self-regulating process, best load rejection may be achieved with I set equal to αD, but this is accidental rather than intentional. On first-order processes, the controller and process deadtimes must match, but when applied to second-order processes, controller deadtime will be higher, and integral time will increase well above αD, as required for minimum-IAE tuning. This approach contrasts markedly with general model-based philosophy, which emphasizes model matching—even for very complex processes, with a single tuning adjustment per loop. The use of all PIDτ_d parameters for tuning does increase the tuning burden to be sure, but it allows a very high level of performance to be reached on any type of process using the same controller.

An insight into its performance can be attained by comparing its load rejection against PI and PID controllers on a process consisting of three noninteracting lags of 10 s each. This process contains no deadtime yet is difficult for the other controllers to regulate. The first load-response curve in Fig. 5.9 is under PI control, having an IAE of 408%-s and a settling time of almost 200 s; in the second, a PID controller has reduced the IAE to 98%-s and the settling time to about 80 s. In the third, the PIDτ_d controller has reduced IAE to 9%-s and the settling time to about 20 s. Improvements this dramatic will not be obtained on higher-order processes or those containing deadtime, yet the power of the controller here is seen not to consist in matching the process but, in fact, in producing phase lead.

The controller described in Fig. 5.8 is fashioned from an interacting PID structure. It is possible to create a noninteracting PIDτ_d controller as well by means of the modification described in Eq. 4.17, whereby an interacting PID controller was converted to a noninteracting version. The resulting device will offer essentially the same marginal improvement as the noninteracting PID controller has over the interacting controller, and again more pronounced in the presence of a secondary lag.

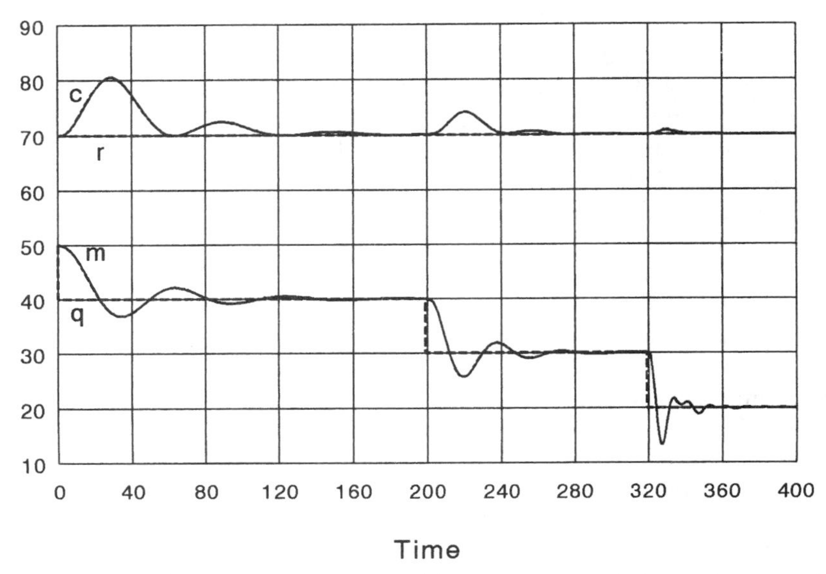

Figure 5.9 Load responses are compared for a process consisting of three equal 10-second lags under PI, interacting PID, and PIDτ_d control, respectively.

Estimating the integrated error sustained by these controllers is similar to the method already derived for PID controllers, with the integral time augmented by the controller deadtime and any filtering:

$$\text{IE}/\Delta m = \mp\frac{P}{100}(I + \tau_d + \tau_f) \tag{5.22}$$

where τ_f is the time constant of the filter located in the optimal position within the controller as described below; no correction is needed for the sample interval.

Phase and gain

The exceptional performance of the deadtime controller, as well as its robustness limitations, can best be observed by comparing the integral mode of the controller with and without deadtime. Figure 5.10 plots the frequency response of a PI controller and a Pτ_d controller; in the former, integration is achieved by positive feedback of the output through a first-order lag, while the latter uses a deadtime element instead.

In response to low frequencies, i.e., approaching steady state, both controllers have the same behavior: gain varying inversely with frequency (increasing directly with period) and phase lag approaching 90 degrees. The PI controller has a dynamic gain approaching 1.0 at higher frequencies and a phase lag gradually approaching zero. By

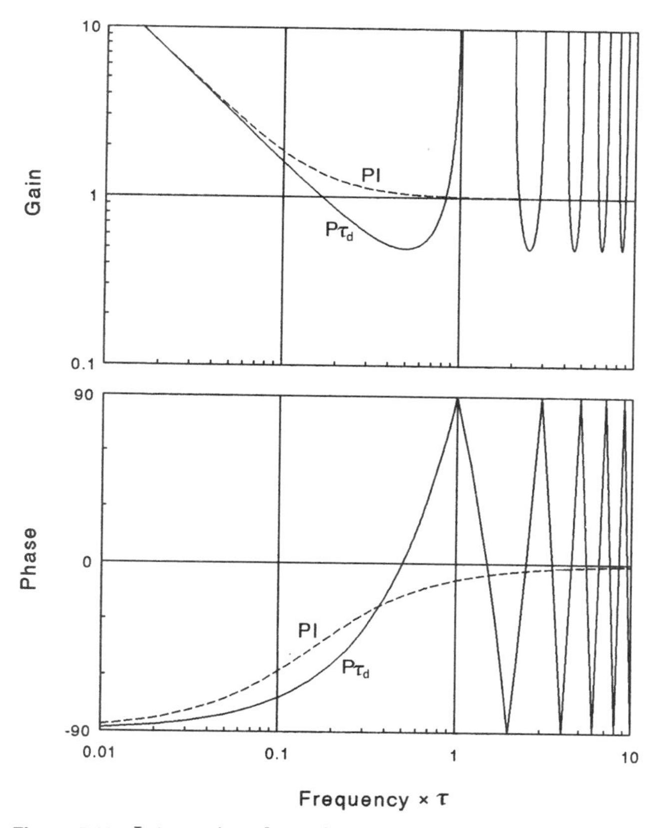

Figure 5.10 Integration through a deadtime block can produce half the dynamic gain as integration through a lag, with phase lead as well.

contrast, the dynamic gain of the $P\tau_d$ controller actually falls to 0.5, and its phase angle crosses zero at a frequency of $0.5/\tau_d$ (period of $2\tau_d$), even contributing a phase lead up to 90 degrees beyond that frequency! The customary penalty of a phase lag associated with integrating action does not have to be accepted when accomplished through a deadtime block. The $PI\tau_d$ controller, when properly tuned, will operate with zero phase lag at the loop period, and the $PID\tau_d$ controller can produce over 80 degrees of phase lead.

With this improved performance potential, however, there is also more risk. For frequencies beyond $0.5/\tau_d$ there lies a range of peaks varying in gain from 0.5 to infinity and phase angles from +90 to −90 degrees. Should operation drift toward this region, high-frequency oscillations can break out easily and expand as well. The number of the harmonics that actually take part in such instability depends on the presence of filtering in the controller and the process. When controlling a deadtime process with no natural filtering, very sharp pulses can erupt; for first-order processes, the same is true if derivative control is used to cancel the process lag. For second- and higher-order processes, however, the higher frequencies are effectively filtered, and tuning is less critical.

Optimal operation requires that the loop frequency or period fall precisely in the first notch of the controller dynamic gain. Because the gain in that notch is only half that of a PI or PID controller, the proportional gain of the PIDτ_d controller can be effectively doubled for a given stability. Therefore, with proper tuning, this controller can provide both lower IE and heavier damping at the same time. However, the notch is quite narrow, and instability can develop if deadtime is either too short or too long. This same property is common to all model-based control loops, in that the function whose frequency response is described in Fig. 5.10 is

$$\frac{m(s)}{e(s)} = \frac{1}{1 - e^{-\tau_d s}} \tag{5.23}$$

which term has appeared in every controller transfer function derived in this chapter.

The PIDτ_d controller does differ from the others in set-point response, however, in that the dominant lead term Ds is not applied to the set point. Therefore, the set-point response has no tendency to ring as in the Dahlin controller, yet overshoot is avoided without resort to set-point filtering, in contrast to the PID controller. Figure 5.11 shows both the load and set-point responses for a PIDτ_d controller on a first-order non-self-regulating process using the optimal load-response settings.

Optimal filtering

This controller needs filtering as much as the others do, and the integral time constant is in the wrong place to do it all. In tests on pure deadtime processes, any mismatch between process and controller deadtimes produces an expanding train of pulses whose width is their difference. Increasing the integral time from zero can attenuate the pulses passing through the positive-feedback loop in the controller, but those passing through the negative-feedback loop from the process remain unattenuated. As a result, filtering using I is not completely effective.

The best place for a filter is a location that is common to both feedback loops. The only place in Fig. 5.8 that fits this description is in the controller output path between the summing junction and the feed-

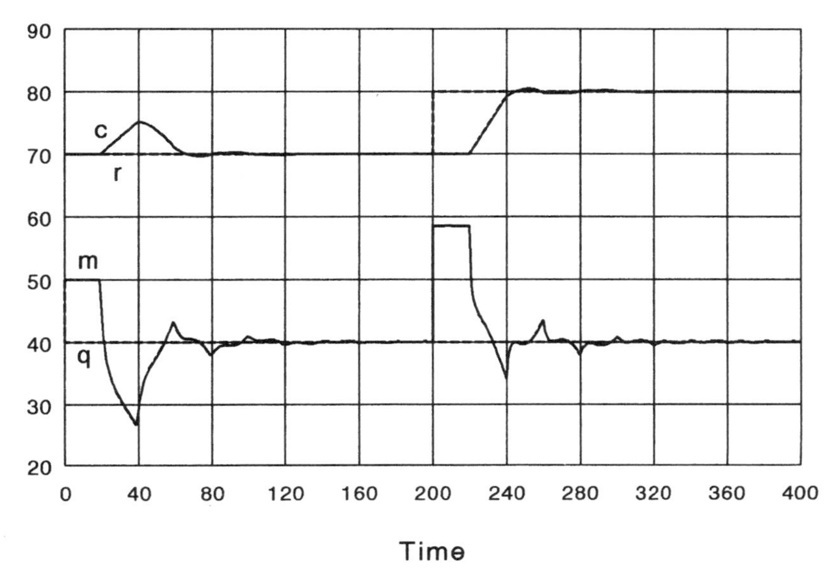

Figure 5.11 Applied to a non-self-regulating process, the PIDτ_d controller produces load and set-point responses superior to those of PID controllers.

back connection to the deadtime block. A first-order filter located there is quite effective in controlling the noise level in the controller output, whether it has risen from a deadtime mismatch or simply from noise on the controlled-variable signal. A second-order Butterworth filter also was tried but did not seem to be as effective in this role.

The optimal amount of filtering applied to a control loop is as little as possible. Because filtering opposes control effort, it should only be used to the extent necessary in preserving stability. In model-based loops, the regenerative nature of the high-frequency oscillations means that filtering *must* be provided—it is not optional, as in the case where noise amplitude is limited. As a result, adaptive filtering is quite effective. Filtering can be automatically increased to quench a growing oscillation and then removed or reduced once a satisfactorily low level of variation has been attained. However, it must not be allowed to increase without limit, since it could defeat control altogether. Limiting its time constant to one-tenth the controller deadtime, for example, can remove this danger.

Performance of Model-Based Controllers

The performance of model-based controllers in response to set-point changes is limited only by the amount of filtering applied and by any saturation that may be encountered. The only differences to be found are caused by lead action in the set-point path or lack of it. This lead caused ringing with the Dahlin controller and in this respect is not trivial. Furthermore, lead causes saturation on even small changes, and saturation may pose other problems if not properly treated. Saturation effects are covered extensively in Chap. 9. However, in the presence of filtering, move suppression, or rate limiting, set-point response generally can be made satisfactory.

The principal performance issue is load response relative to the best practical as a norm. Model-based controllers are split into two camps on this front. The first camp uses only model-to-process matching, with a filter as the only tuning adjustment, such as, for example, IMC without an analytical predictor, or a matched Smith predictor. The second camp attempts to tune for improved load response by whatever means available, possible with the Dahlin and PIDτ_d controllers. Their respective results depend on the dynamics of the controlled process.

First-order processes

For a pure deadtime process, any of the preceding methods gives best possible load and set-point response with the model matched to the process. Controller output is stepped to the new steady-state value at the instant a deviation appears and then remains there. As a result, the deviation is sustained for one deadtime only, at the end of which it returns to zero and remains there. IE is the product of the load step, process gain, and deadtime for load changes and simply the set-point step and deadtime for set-point changes.

Differences between what is possible and what a controller actually achieves begin to appear as a time lag enters the loop. If the controller model continues to be matched to the process, a step-load change will still produce an equal step in controller output one deadtime later, giving the same IE as above. However, this is no longer the best possible nor even the best practical response. The best practical IE varies exponentially with the deadtime-to-time constant ratio, as shown in Eq. 5.6. This practical performance relationship is plotted against the ratio of deadtime to total process response time, i.e., $\Sigma\tau = \tau_d + \tau_1$, in Fig. 5.12 and labeled IMC. On this scale, the pure-deadtime process lies at 1.0 and the non-self-regulating process at 0.

The values of IE/IE$_b$ for the other controllers described in Fig. 5.12 were all determined by simulation, with controllers tuned for minimum IAE in response to step-load changes introduced at the controller output. The two PID controllers and the PIDτ_d controller had their derivative filters fixed at $D/10$. Performance of the PIDτ_d con-

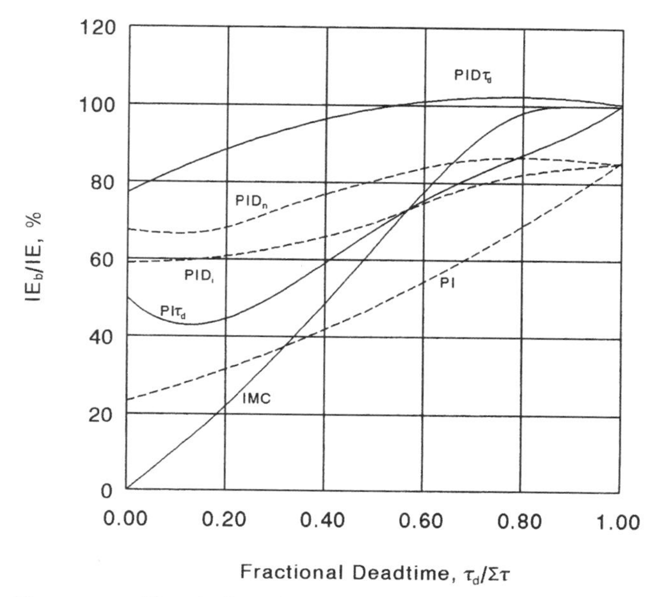

Figure 5.12 Practical performance ratings for controllers on first-order processes.

troller improves markedly with increasing derivative gain, although any increase is difficult to justify on first-order processes, where there is little natural filtering. The performance of the IMC controller in Fig. 5.12 is not handicapped by any filtering, which must be present in any real application.

There are two points of convergence on the right axis, the upper at 100 percent performance for model-based controllers and the lower for PID controllers, where derivative is set to zero as being ineffective against pure deadtime. Performance of the PI controller falls uniformly as the process moves toward lag dominance but not as rapidly as does that of the IMC controller. Performance of the two PID controllers varies little across the process spectrum, both being better than the IMC controller for all lag-dominant processes. The PIτ_d controller surprises by performing almost as well as the interacting PID controller for lag-dominant processes and better for deadtime-dominant processes.

Recognize that the superiority of the PIDτ_d controller lies in its tuning. If it were simply employed in a matching role, it would perform no better than the IMC controller, which it resembles functionally. Interestingly, it is able to produce a lower IE than the best practical on deadtime-dominant processes, as indicated by performance ratings slightly exceeding 100 percent. The Dahlin controller is also tunable and could approach this level of performance on first-order processes given fast sampling and minimum filtering; its set-point response is hyperactive, however, where that of the PIDτ_d controller is not. I also have been able to tune the modified Smith predictor to produce similar results,[12] and if its controller gain equals the derivative gain of the PIDτ_d controller, load responses will be identical; it is much more difficult to tune, however.

Higher-order processes

As covered in Chap. 2, IE_b is not affected by the presence of secondary lags, although they certainly make control more difficult. Higher-order controllers than described above are possible, but tuning them to maximize performance is not really practical, there being too many degrees of freedom relative to the observable characteristics of a typical response curve. The most common multiorder controllers are matrix-based, and they are tunable only by selecting the control horizon and move suppression. Neither are they really capable of minimizing IAE or IE, as is possible using all available parameters on a PIDτ_d controller.

To illustrate the problem posed by secondary lags, Fig. 5.13 compares the practical performance of several controllers as a function of the ratio of τ_2/τ_d for a non-self-regulating process. The left-hand edge

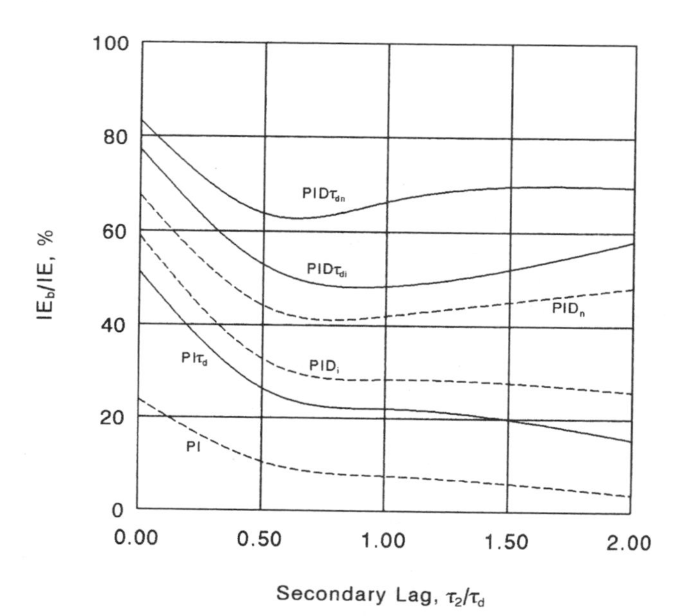

Figure 5.13 Practical performance ratings for controllers on second-order non-self-regulating processes.

of this plot, where τ_2 is zero, is coincident with the left-hand edge of Fig. 5.12. The performance of the PI controller is low on non-self-regulating processes and only goes lower as τ_2 increases. The PIτ_d controller starts higher but similarly declines with increasing τ_2. The two PID controllers both suffer declining performance, but the noninteracting version begins to recover where τ_2 exceeds τ_d, while the interacting controller does not. This was explained at the end of Chap. 3 to be caused by the inability to set its effective values of I and D close enough together.

Two PIDτ_d controllers also were tested on these second-order processes with similar results. The noninteracting version has a higher performance across the range of τ_2, and the performance difference increases above a value equal to τ_d. However, the interacting version does show some recovery in this area, continuing to outperform the noninteracting PID controller.

All the controllers with derivative action had filters with $\alpha = 0.1$. Not shown in the figure are two test results for the PIDτ_d controllers where α was lowered to 0.02. This raised performance at τ_2/τ_d of 1.0 for the interacting version from 48 to 58 percent and that of the noninteracting version from 66 to 73 percent. The ability to adjust the derivative filtering in these controllers consistent with the noise level on the controlled variable allows the possibility of significant performance enhancement on high-order processes. Some retuning is required following such an adjustment, however, since a reduction in filtering allows a reduction in controller deadtime. This relationship obviously does not apply to first-order processes, where the controller and process deadtimes must match. (It may be preferable to have the derivative filter in this controller linked to the deadtime setting rather than to the derivative setting.)

Effect of the sample interval

Although not a tuning parameter, the selection of sample interval for model-based controllers has a pronounced effect on performance. All the performance ratings described above apply to fast sampling only, i.e., where the deadtime is 20 sample intervals or more. In the last chapter, performance of PID controllers was seen to be affected more than PI controllers by the sample interval due principally to its impact on derivative gain. With model-based controllers, the impact is even greater.

The same lag-dominant process ($\tau_d = 20$, $\tau_1 = 100$) controlled by a PID controller in Fig. 4.7 is controlled by an interacting PIDτ_d controller in Fig. 5.14. When sampled at $\tau_d/20$, IE/Δm is 21.4 compared with 30.2 for the interacting PID controller. However, increasing the sampling interval to $\tau_d/2$ causes IE/Δm to increase to 47.2, *more than*

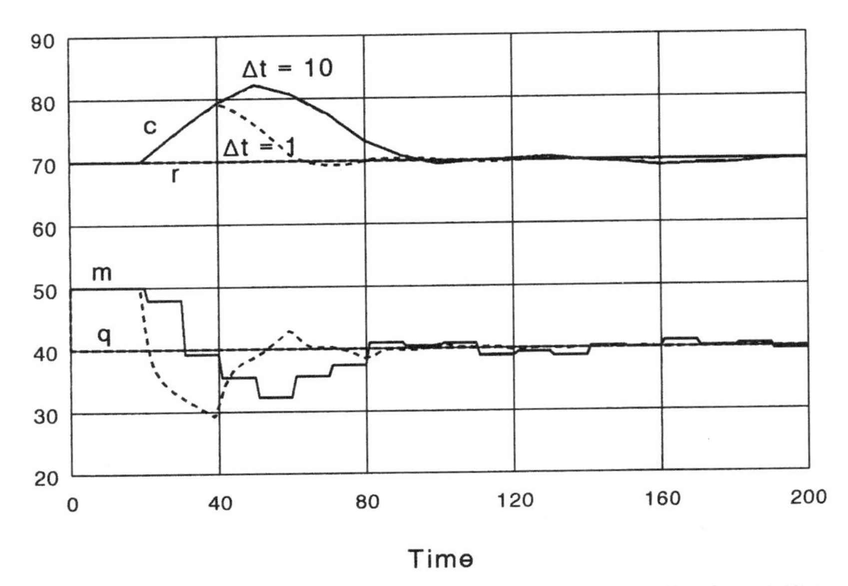

Figure 5.14 Increasing the sampling interval of a PIDτ_d controller from 0.05 to 0.5 deadtime loses all the performance advantages of deadtime compensation.

double, and over 50 percent higher than the fast-sampling PID controller! In other words, any performance advantage gained by deadtime compensation is readily lost by slow sampling. While this rule obviously does not apply to deadtime processes where the sample interval can be set equal to the deadtime, it is of paramount importance with lag-dominant processes.

Obviously affected are matrix-based controllers that are computationally intensive, therefore requiring long sample intervals. The process modeled in Fig. 5.7 is a case in point. With a τ_1/τ_d ratio of 5, allocation of 30 samples to three time constants results in a sampling interval of $\tau_d/2$, with the resultant performance deterioration shown in Fig. 5.14. This combination of dynamics is quite common in multistage and distributed processes such as distillation columns, heat exchangers, and reactors.

Setting the sample interval of a controller to one-tenth the dominant lag is frequently recommended in process control. While this practice may produce satisfactory results on deadtime-dominant processes, it certainly does not when applied to lag-dominant processes. A preferred recommendation would have the sample interval set at one-tenth the process *deadtime*—a practice that would apply equitably to all processes. However, this would place a severe burden on matrix-based controllers, requiring 200 or more coefficients to model a lag-dominant process satisfactorily.

Robustness of Model-Based Controllers

There are some important differences between PID and model-based controllers from the standpoint of robustness. First and foremost, there is a general rule that high performance in a controller necessarily diminishes robustness. Since model-based controllers have—or can have—higher performance than PID controllers, they can be expected to be less robust, and they are. However, a given model-based controller does not have to have a high performance just because it is not robust. In other words, high performance brings low robustness with it, but the reverse is not necessarily true. An IMC controller applied to a lag-dominant process, for example, could have the low performance shown in Fig. 5.12, and a low robustness as well—the worst of both worlds.

As described in Chap. 2, robustness can be measured as the fractional departure of any process parameter from its initial value that would cause the control loop to reach its limit of stability. The principal process parameters likely to change are gain and deadtime; these are also the most sensitive parameters in a model-based controller with respect to their effect on stability.

Gain margins

In a PID loop, a uniform oscillation results when the loop gain—the product of all its steady-state and dynamic gains—reaches unity at the period of the oscillation. This relationship has been used successfully to tune controllers, a topic to be investigated in the next two chapters. With model-based control, however, multiple loops exist, even when there is but a single controller and controlled variable. In the simplest case, that of IMC in Fig. 5.1, there are two feedback loops—one negative and one positive—that are intended to cancel. This much is common to all model-based control, although, as in the modified Smith predictor of Fig. 5.6, as many as four loops exist.

If the two loops in the IMC controller do cancel because of a perfect match between the model and the process, the loop gain is actually zero for all frequencies. The result of this cancellation is *deadbeat* response: For a single-step upset, the controller makes but one move, resulting in neither overshoot nor oscillation of the controlled variable. The analysis is simple for a deadtime process. A step change in set point moves the controller output in a step; one deadtime later the controlled variable responds an equal amount, because the gain of the negative-feedback loop is 1, the controller being the inverse of the process (except for its deadtime). This unit gain for the negative-feedback loop is required to drive c to match r in one move, yet oscillation is avoided by cancellation through the model in the positive-feedback loop. All model-based control operates essentially the same way, in that a negative-feedback loop gain of unity really amounts to an overall loop gain of zero.

Next, consider a simple mismatch of static gains. If the static gain product $\prod K$ is less than 1, an insufficient correction will be made in one pass, resulting in an exponential approach to the new steady state. If $\prod K$ is 0.5, for example, the new set point will be approached halfway each deadtime, producing an exponentially diminishing staircase. For $\prod K$ between 1 and 2, overshoot will result, producing a damped oscillation having a period of two deadtimes; the decay ratio will be $\prod K - 1$, measured as the ratio of the differences in height between two successive peaks and their common valley. When $\prod K$ reaches 2.0, the correction is double the required amount, resulting in 100 percent overshoot, followed by 100 percent overshoot, etc.

Thus a loop gain of 2 in model-based control produces uniform oscillation just like a loop gain of 1 in PID control. The robustness of the typical model-based loop in the gain dimension is exactly 100 percent, in that a 100 percent increase in process gain from the initial condition will produce an undamped cycle. This compares with a robustness of 97 percent for the PI controller applied to the deadtime process in Fig. 2.13.

For processes containing lags, the IMC controller produces the inverse of the process dynamic gain vector, so the preceding analogy remains true of the static gain product. However, a higher performance is attained by increasing the steady-state loop gain above 1, forcing the manipulated variable to overshoot a load change as in Fig. 5.2. This higher gain under optimal tuning means a lower gain margin between that point and the limit of stability. PI and interacting PID and PIDτ_d controllers were tested on a first-order non-self-regulating process and found to have gain margins of 48, 40, and 30 percent, respectively, between optimal load response and instability. (Process gain was increased by decreasing the integrating time constant to the point where uniform oscillations were produced.) Reducing the derivative filter factor α in the PIDτ_d controller from 0.1 to 0.02 had very little effect on either gain margin or performance. When a secondary lag equal to the deadtime was added to the process, the gain margins did not change appreciably except for the PIDτ_d controller with reduced α, whose gain margin decreased from 30 to 20 percent, which correlates with the 10 percent increase in performance reported above.

Deadtime margin

The substantial cushion in loop gain inherent with matched-model controllers is not there for deadtime variations. Again, the pure deadtime process is the clearest example. With perfect matching, a step in the manipulated variable produces identical steps in the outputs of

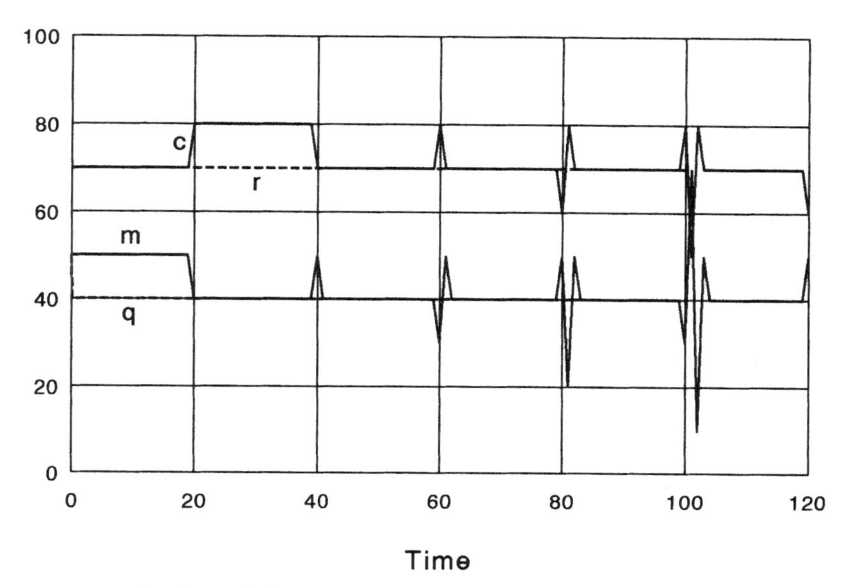

Figure 5.15 A mismatch between process and controller deadtimes produces an expanding pulse train in the absence of filtering.

process and model that cancel. However, the slightest difference in timing between the arrivals of the two steps creates a pulse whose height is that of the step and whose width is the deadtime difference. Unfortunately, the story does not end here, for without any lags in either feedback loop, the residual pulse continues to circulate through both loops, replicating itself with each passage, as shown in Fig. 5.15. The result is a quickly expanding oscillation. In the figure, the controller deadtime is one sample interval longer than the process deadtime, so the first output pulse is in a direction opposite to the upset. If the controller deadtime were the shorter of the two, the pulse would develop in the direction of the upset, but the accumulation of pulses appears the same in either case. Incidentally, a step upset is not required to produce a growing pulse train—the tiniest disturbance, even white noise, will serve as well.

With both the process and model simulated in the same computer, a match is easy to achieve, but this is a contrived situation. In practice, the process and model will be driven by different clocks, so perfect matching is impossible to achieve. Connecting a real deadtime controller to a simulated deadtime process through a communication link will demonstrate this point. Deadtime mismatches are therefore inevitable.

All deadtime models are digital, being simulated by storing a finite number of data points in a register and either moving them from address to address each sample interval or moving input and output pointers from address to address. Deadtime is therefore an integer number of sample intervals, and mismatch is measured in an integer number of sample intervals. The smallest mismatch that produces the result shown in Fig. 5.15 is a single sample interval. The deadtime margin to instability in a model-based control loop is consequently $+0$, $-\Delta t$. Increasing the sample interval relative to the deadtime therefore increases the robustness of a loop to deadtime variations. The deadtime margin expressed in percent is $+0$, $-100\Delta t/\tau_d$; it is important to remember that a mismatch is harmful in *either* direction, in contrast to PID loops, where instability will result only upon an increase in process deadtime. For example, the PI loop whose stability limit appears in Fig. 2.13 shows that deadtime must increase 130 percent to reach it, and the envelope is open in the other direction.

The performance-robustness tradeoff was raised in Chap. 4 with regard to the application of the sampled integral controller to a deadtime process. The sample interval had to be equal to or longer than the process deadtime for robustness. The performance penalty in that case was shown in Fig. 4.8 to be caused by uncertainty in load response over the sample interval. That controller was essentially a model-based controller having a deadtime register containing only one sample. The use of multiple samples shortens the sample interval and therefore reduces the uncertainty, but it also lowers the robustness by the same degree.

A second approach to improving robustness is the addition of filtering. The filter was the only tuning parameter in the IMC controller, whereas in the Smith predictor, reducing the controller proportional gain from its maximum produced filtering. In a private communication, Professor Richalet, inventor of the model-predictive controller IDCOM, admitted to reducing the controller gain by a factor of 4 to 6 below the maximum to ensure a robust loop. For a deadtime process, adding a first-order filter of 10 percent of the deadtime moves the limit of stability for deadtime variations to ± 20 percent while increasing IE by the 10 percent size of the filter. Filtering consequently produces a favorable tradeoff when there is no natural filtering in the loop.

Figure 5.16 shows the effect of an adaptive filter in the presence of a 5 percent mismatch in deadtimes. Its time constant begins at zero prior to the first load change, adapting as required to prevent the pulse train from growing. When the second load step arrives, it has remained at a level of 2 percent of the controller deadtime, which shows in the recovery curve but also limits the size of the first pulse. The following set-point response is similarly affected.

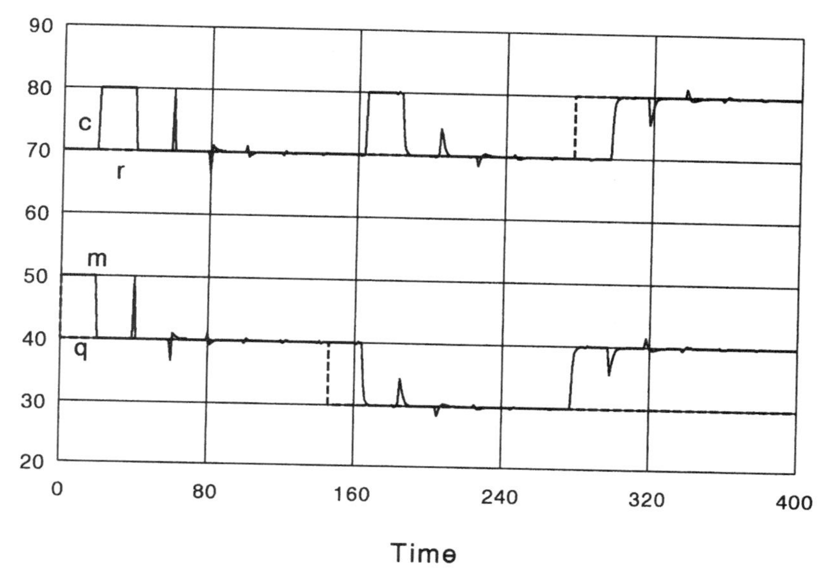

Figure 5.16 An adaptive filter can quickly tune itself to accommodate a 5 percent mismatch between process and controller deadtimes.

For first-order processes, if the controller has a lead that effectively cancels the process lag, robustness to deadtime variations is the same as for a deadtime process. With increasing ratios of τ_1/τ_d, however, the size of the filter accompanying the controller lead grows with respect to the deadtime as well. This is true at least of the PIDτ_d controller having a fixed derivative filter factor α. Consequently, lag-dominant first-order processes appear to be more robust in the face of deadtime variations than do deadtime-dominant processes. In the case of the Smith predictor or IDCOM, however, where the filter is the model lag divided by the controller gain, maximum gain produces the same minimum robustness observed with deadtime processes, but reducing the gain only slightly can give very effective filtering.

Figure 5.17 shows what happens to the lag-dominant first-order process controlled by PIDτ_d in Fig. 5.14 when process deadtime is reduced by 10 percent. The controller sampling at $\tau_d/2$ remains stable, while that sampling at $\tau_d/20$ reaches its limit of stability. Of course, the fast-sampling PID controller also would remain stable in the same situation and continues to outperform the slow-sampling PIDτ_d controller.

For first-order non-self-regulating processes, optimally tuned PI and PID controllers both reach a stability limit upon an increase in deadtime of 38 percent, somewhat below their gain margins reported above. Adding a secondary lag to the process makes very little differ-

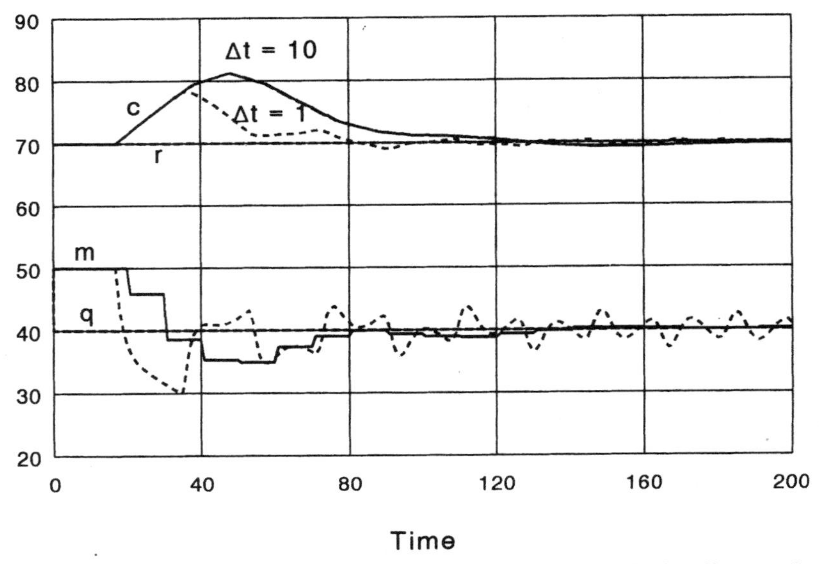

Figure 5.17 A 10 percent reduction in process deadtime has little effect on the PIDτ_d controller with slow sampling but destabilizes the one with fast sampling.

ence in robustness. For the same first-order process with a PIDτ_d controller sampling at $\tau_d/20$, stability limits are reached with a 20 percent increase in process deadtime and a 10 percent decrease. Adding a secondary process lag equal to the deadtime makes the deadtime margins equal at ± 25 percent. Reducing α from 0.1 to 0.02 then reduces the margins to ± 20 percent.

Robustness versus performance

The tradeoff between robustness and performance can be displayed graphically both as a function of controller selection and of tuning for any given process. Just one example is given here, that of the lag-dominant first-order process having a τ_1/τ_d ratio of 5.

Seven controllers are compared in Fig. 5.18. Of these, the IMC controller has both the lowest performance and the lowest robustness at the same time; identical behavior would be obtained using a matched Smith predictor or Dahlin controller. Performance is low because it is measured in response to a load change entering upstream of the process lag. Robustness is 5 percent due to sampling at $\tau_d/20$ without filtering. The same sampling rate was used for all the other controllers except for the slow PIDτ_d controller shown in Figs. 5.14 and 5.17, sampling at $\tau_d/2$. The fast PIDτ_d controller shows the best performance of all, but remember that the Smith predictor could duplicate it if properly tuned, and the same is true of the Dahlin controller.

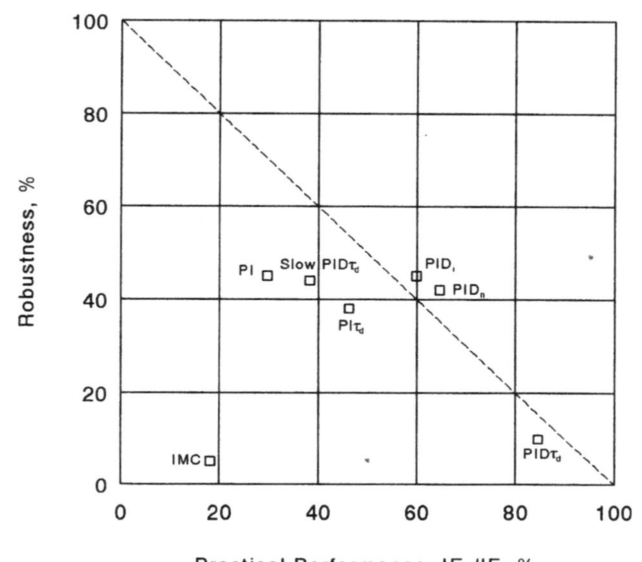

Figure 5.18 A comparison of several controllers applied to a lag-dominant process on a scale of robustness versus performance.

The pattern of the points would be expected to fall in a diagonal connecting 100, 0 with 0, 100, with the faster PIDτ_d controller and the two PID controllers almost lying on it. Note that the relationship of the PID controllers to one another parallels the diagonal. Also, the impact of the sample interval on the PIDτ_d controller roughly follows the diagonal. In essentially the same way, detuning a controller to improve robustness at the expense of performance will follow a diagonal trend.

On a deadtime process, an infinitely fast deadtime controller will have a performance of 100 and a robustness of 0. Adding a filter was observed to improve robustness more than it sacrificed performance, placing its point above the diagonal—a worthwhile tradeoff. On the same process, however, a PI controller would show a robustness of 97 percent at a performance rating of 80 percent, as noted in Chap. 3. This point falls well above the diagonal. Therefore, each type of process will tend to have its own distribution of points for the various controllers.

No weighting factors have been applied to Fig. 5.18, and in a given situation, they may be required. For example, with a process known to be stationary, performance may be weighed much more heavily than robustness. Conversely, for a loop operating on a remote and unattended plant, robustness may be more important.

Another factor worth considering, however, is the use and effectiveness of self-tuning. If a controller can be tuned and retuned automatically on-line, then the robustness issue may be unimportant, and full emphasis can be placed on performance, providing, of course, that the self-tuner is itself robust.

Notation

a_i	Step-response coefficient at sample i
A,B	Exponential operators for Dahlin controller
c	Controlled variable
$\hat{c}$	Estimate of controlled variable
d	Differential operator
D	Derivative time
DMC	Dynamic matrix control
e	Deviation, 2.718
F	Filter
G_c	Controller
G_m	Model
G_m^*	Model without deadtime
G_p	Process
G_q	Load dynamics
G_q^*	Load dynamics without deadtime
I	Integral time
IAE	Integrated absolute error
IE	Integrated error
IE_b	Best practical IE
IMC	Internal model control
K	Proportional gain of Dahlin controller
K_c	Proportional gain of controller
K_m	Model steady-state gain
K_p	Process steady-state gain
ΠK	Product of steady-state gains
m	Manipulated variable
MPC	Model-predictive control
n	Number of samples
N	Number of samples in the deadtime
P	Proportional band
q	Load variable
r	Set point
s	LaPlace operator d/dt

t	Time
Δt	Sample interval
x	Variable
α	Ratio of derivative filter time to derivative time
Π	Product
Σ	Sum
τ_d	Deadtime
τ_f	Filter time constant
τ_1	Primary time constant
τ_2	Secondary time constant

Subscripts

C	Number of samples to control horizon
n	New
P	Number of samples to prediction horizon

References

1. Smith, O. J. M., "A Controller to Overcome Dead Time," *ISA J.*, February 1959.
2. Lipták, B. G., *Instrument Engineer's Handbook,* Vol. 2, *Process Control,* Chilton, Philadelphia, 1970.
3. Lipták, B. G., *Instrument Engineer's Handbook,* rev. ed., Chilton, Radnor, Pa., 1985.
4. Dahlin, E. B., "Designing and Tuning Digital Controllers," *Instr. Control Sys.* 41(6), June 1968.
5. Richalet, J. A., Rault, A., Testud, J. L., and Papon, J., "Model Predictive Heuristic Control: Applications to an Industrial Process," *Automatica,* 14:413–428, 1978.
6. Cutler, C. R., and Ramaker, B. L., Dynamic Matrix Control—A Computer Control Algorithm, AIChE National Meeting, Houston, Texas, 1979.
7. Seborg, D. E., Edgar, T. F., and Mellichamp, D. A., *Process Dynamics and Control,* Wiley, New York, 1989, pp. 278–280.
8. Morari, M., and Lee, J. H., "Model Predictive Control: The Good, the Bad, and the Ugly," *Chemical Process Control—CPCIV,* AIChE, New York, 1991, pp. 429–431.
9. Doss, J. E., and Moore, C. F., "The Discrete Analytical Predictor—A Generalized Dead-Time Compensation Technique," *ISA Trans.,* 20(4): 77, 1982.
10. Seborg, *op. cit.,* pp. 639–645.
11. *Ibid.,* pp. 629–631.
12. Shinskey, F. G., "Putting Controllers to the Test," *Chem. Eng.,* December 1990, pp. 96–104.
13. Seborg, *op. cit.,* pp. 661, 662.
14. Cutler, C. R., and Johnston, C. R., "Comparison of the Quality Criterion for PID and Predictive Controllers," *Proceeding of the American Control Conference,* Boston, 1985, pp. 214–219.
15. Luyben, W. L., *Practical Distillation Control,* Van Nostrand Reinhold, New York, 1992, pp. 262–267.

Part

3

Controller Tuning

Chapter

6

Manual Tuning Methods

Recent developments in model-based control have attempted to eliminate the problem of controller tuning or at least reduce it to a single adjustment such as filtering or move suppression. However, as amply demonstrated in the last chapter, tuning a model-based controller specifically for load rejection can produce a much higher performance than simple model matching. While tuning has been considered mostly an art, it should not become a lost art. Ziegler and Nichols were perhaps the first who attempted to reduce it to a science, with some success, since their methods are still used widely after 50 years.

The principal limitation of their methods is that they were developed using a single type of controller on a single type of process for a single type of disturbance. While their choices of controller, process, and disturbance were excellent in that they covered the most common combination likely to be found among fluid processes, their work was far from comprehensive. In this chapter, the science is made more exact through computer simulation and more comprehensive by investigating other controllers, processes, and disturbances.

Successful tuning requires two efforts: process identification and application of the appropriate tuning rules. The rules presented in this chapter, while complex in order to be comprehensive, are relatively easy to apply. The difficult part is identifying the process with sufficient accuracy so that the tuning rules will in fact result in a maximum level of controller performance. And, of course, the identification problem exists even if one intends not to tune but to match the controller model to the process. In college textbooks, the student is always "given" a set of process parameters and asked to "design" a controller to reach a "desired" level of performance. In a process plant, however, the control engineer is given *no information* and expected to produce as high a level of control performance as is possible. Tuning begins with identifying the process.

Open-Loop Process Identification

To identify an unknown process, its response to known disturbances must be analyzed. This can be done either in the open or the closed loop. In the open loop, no control is being applied, so the process will not necessarily be in a settled initial or final condition—particularly a problem with non-self-regulating processes. Given enough manual regulation to bring about an initial steady state, however, the process may be classified, and even its steady-state gain may be determined.

In the closed loop, some regulation is applied by a controller, preventing the process from wandering from the limits of the controller's proportional band. However, the information gleaned from closed-loop testing is focused in the band around the natural period of the process. This is not enough to classify a process, but it still can be used to tune a controller satisfactorily. Combining the two methods will enable the most accurate tuning.

There is a reluctance among plant management to permit disturbances to be introduced for purposes of controller tuning. Yet, unless known disturbances are applied, the information required for effective tuning will not be obtained, and controller performance will suffer. If the disturbances are limited to a very small amplitude, comparable with the noise level on the controlled variable, then many disturbances and considerable time may be required to provide accurate enough information to tune the controller. By contrast, one large disturbance can provide the same information in the minimum time. However, each type of disturbance also has its advantages and disadvantages, as described below.

Sinusoidal testing

Beginning from a steady state, a disturbance may be introduced into the controller output to elicit a response in the controlled variable. Step, pulse, ramp, and sinusoidal disturbances have all been used. Because sinusoidal disturbances contain a single frequency, the information produced from a single test—phase and gain—is not very useful by itself. Several different frequencies must be tried over a spectrum where the process phase lag varies from 0 to beyond 180 degrees. These tests tend to be very time-consuming, with such a low return on investment that they are rarely used on fluid processes. For electronic and mechanical systems such as filters, controllers, motors, valves, and instruments, however, the tests can be completed much faster and therefore provide useful information.

The most useful area of the frequency spectrum is where 180 degrees of process phase lag is produced. This is known as the *natural frequency* or *period,* since an undamped loop will oscillate here if the controller has no phase lag or lead. This would be the case if a proportional or optimally tuned PIτ_d controller were applied. However, if the process is unknown, the natural period would have to be guessed, a sinusoidal wave introduced at that period, and the resulting phase lag measured. Other periods would then have to be tried if the first guess was not close. It is far easier and faster to identify the process dynamics at that point using a closed-loop method.

Frequency-response testing has other limitations as well. It does not reveal nonlinearities, nor is it particularly useful in determining steady-state gains. While promoted in the 1950s, it has never been used widely in process control.

Step responses

Step disturbances have the advantage of containing all frequencies in one easily reproduced event and therefore have been the most commonly applied test over the years. A typical response of the controlled variable to a step disturbance in the manipulated variable is the "reaction curve" of Fig. 6.1, so called by Ziegler and Nichols.[1] They only

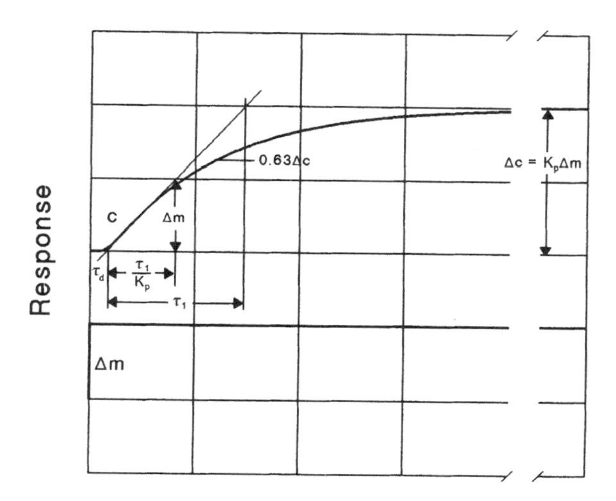

Figure 6.1 In a second-order step response, τ_1 estimated from the maximum slope tends to be longer than the true value obtained from the time required for a 63 percent response.

used two pieces of information from the curve—its deadtime and maximum slope—to apply their tuning rules. The curve shown represents the response of a self-regulating process. If the process were non-self-regulating, there would be no final steady state, since the slope would continue indefinitely. In either case, the maximum slope is useful in estimating the required proportional gain of the controller.

The deadtime is estimated as the point of intersection of the baseline with the extension of the maximum slope. Following the elapse of this deadtime, the time required for the controlled variable c to change an amount equal to the step input Δm following the maximum slope would be the time constant τ_1 for a non-self-regulating process and τ_1/K_p for a first-order self-regulating process. If the test is continued until a new steady state is reached, then the steady-state gain K_p can be estimated by dividing the steady-state change Δc by the size of the step. Having done that, τ_1 may then be determined for the self-regulating process. However, about five time constants must pass for 99 percent complete response, which is required for an accurate estimate of K_p using this method. For some processes, the wait could be hours, in which case unrequested disturbances could intervene, affecting the results. This can be expected in the response of temperature or composition in a distillation column, where the dominant lag could be an hour or more.

Two other techniques can identify the dominant time constant if the final steady state has been reached. Extending the slope to the final value of the controlled variable will cover a time equal to τ_1 if the process is essentially first-order. The second technique measures the elapsed time required for the controlled variable to change 63.2 percent $(1 - e^{-1})$ of the difference between its initial and final steady states. Any secondary lag, if present, would be lumped with the deadtime. The reaction curve plotted in Fig. 6.1 is actually second-order, with the secondary lag equal to the deadtime at $\tau_1/10$. In the presence of the secondary lag, τ_1 estimated by the slope method and the 63 percent response method do not agree, because the secondary lag reduces the slope at the beginning of the response curve where the first-order curve has its highest slope. Notice the difference between the two results in Fig. 6.1. In practice, the 63 percent point is a more accurate representation of the dominant time constant and should be used in the open-loop tuning rules given in this chapter, which include provision for a secondary lag. However, a step-response curve does not allow the accurate identification of a secondary lag. For controller tuning purposes, the slope is a reasonable indication of how the controller's gain should be set, which explains the use of the slope by Ziegler and Nichols.

Their tuning rules using this method apply principally for step-load changes to non-self-regulating and lag-dominant processes with deadtime. While covering the most problematic loops commonly found in fluid processes, the method left out deadtime-dominant processes and set-point responses. For a self-regulating process, effective tuning does depends on the ratio of τ_d/τ_1, but unless K_p is determined as well, neither is τ_1. Ziegler and Nichols were probably wise not to attempt to extract more than two parameters from a step test, especially with the instruments they had to work with in 1941. Reference 2 describes a graphic method for extracting a secondary lag from a step-response curve, but it requires a large step, a final steady state, and a linear process and fails as τ_2 approaches τ_1.

Another important limitation of the step test is that it is affected by a common valve nonlinearity known as *dead band,* but without being able to identify it. Dead band is caused by friction in the valve stem. To move the valve, enough force must be applied to the stem to overcome this friction. Whenever the direction of the signal is reversed, a similar force must be developed in the opposite direction before the valve stem reverses direction. In effect, the stem moves along different paths for increasing and decreasing signals, stopping in between, as shown in Fig. 6.2. The dead band is identified as a in the figure.

Suppose that the last motion of the controller output and valve stem had both been upward. If an upward step is next introduced into the controller output, the stem will follow it exactly, and a true indication of the process characteristics would be revealed. If, however, a downward step were applied instead, the controller output would have to cross the dead band before the stem would move, reducing the magnitude of the step change by a. Since the slope depends on the

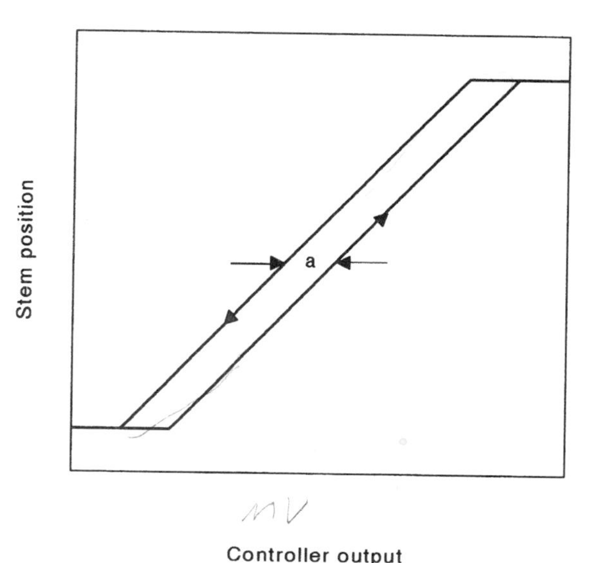

Figure 6.2 Dead band, caused by friction, separates the opening and closing paths of a control valve.

size of the step, estimates of both τ_1 and K_p using the slope would be in error. If the process is self-regulating and allowed to reach a new steady state, τ_1 determined either by the extension of the slope to the new steady state or the point of 63.2 percent response would be accurate, but K_p would not.

To be sure, dead band affects the controller settings and the controller's performance as well and, as a result, should be identified. However, a single step cannot make that identification. To make the identification, a step in one direction may be followed by another step in the same direction and the results compared. If the controlled variable responds the same to both tests, either there is no dead band or the first test started from an initial condition that was also in the same direction. However, if the process or valve were nonlinear, different results could be obtained even with no dead band. If the second step were in the opposite direction so that the output returned to its original position but the controlled variable did not (for a self-regulating process), dead band would be indicated. This last procedure amounts to a pulse test, discussed next.

Dead band always interferes with control, causing limit cycles when integral action is used in the presence of a process integrator.[3] It should be eliminated from product-quality and other important loops through the use of a valve positioner or by manipulating flow with a cascade controller. Dead band can be as high as 10 percent in a worn valve but is usually reduced below 0.5 percent by a valve positioner.

A single pulse

The use of a single pulse disturbs the process ultimately less than a step of the same height. The step left the non-self-regulating process continuing indefinitely on a ramp away from its original steady state; returning m to its original value can at least stop the ramp, in the absence of dead band, as shown in Fig. 6.3. In the presence of dead band, the slope of the ramp will be decreased, but not to zero. If the first step began from a steady state and an initial condition in the same direction, the second step would be smaller by a. The initial slope dc/dt would be $\Delta m_1/\tau_1$, and the final slope would be a/τ_1. For a self-regulating process, the controlled variable will return to its original steady state in the absence of dead band. Any difference between the steady states would equal a/K_p.

For the non-self-regulating process, at the end of the (first) ramp, the controlled variable will have moved a distance

$$\Delta c_1 = \Delta m_1 \tau_w / \tau_1 \tag{6.1}$$

where τ_w is the width of the pulse. This allows the integrating time constant τ_1 to be estimated at the change of slopes if there is dead

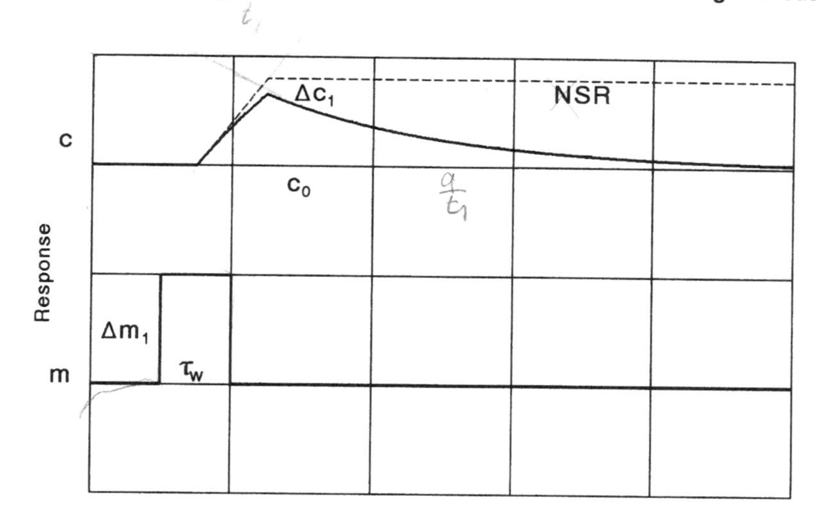

Figure 6.3 A single pulse returns the self-regulating process to its original steady state.

band and no secondary lag or from the change in steady states if there is a secondary lag but no dead band. (Dead band should really be eliminated from a non-self-regulating loop in any case.)

For the self-regulating process, the pulse width will in most cases be too short to allow a new steady state to be reached. Consequently, the time constant and steady-state gain must be estimated by means different from the step test. In response to the pulse, the output of a first-order lag will reach a peak shown in Fig. 6.3 having a value of

$$\Delta c_1 = \Delta m_1 K_p (1 - e^{-\tau_w/\tau_1}) \tag{6.2}$$

While Eq. 6.2 contains both unknowns τ_1 and K_p, the area under the pulse-response curve is only a function of the latter:

$$\int (c - c_0)\, dt = \Delta m_1 \tau_w K_p \tag{6.3}$$

Therefore, integrating the controlled variable over the course of the test (assuming no dead band) can determine the gain, allowing Eq. 6.2 to be solved for the time constant. Recognize, however, that integration must proceed for four to five time constants, presenting the same problem as waiting for the steady state following a step test.

If the process is deadtime-dominant to the extent that the time constant is only a fraction of the pulse width, then the results of the pulse disturbance are essentially the same as that of a step. Then K_p can simply be estimated as $\Delta c_1/\Delta m_1$, and the time constant can be estimated by extending the maximum slope or marking the 63.2 percent response, as shown for the step in Fig. 6.1. The value of any deadtime can be estimated for all processes by extending the maximum slope to the baseline, as done for the step test in Fig. 6.1.

The doublet pulse

By following the first pulse immediately by an equal and opposite pulse as shown in Fig. 6.4, even the non-self-regulating process will be returned to the original steady state (in the absence of dead band). For the self-regulating process, the resulting response curve will be distributed equally on both sides of the original steady state on the basis of area, but the heights of the two peaks produced will differ in most cases and can be used to estimate the time constant independent of the gain.

Using the doublet pulse, even the secondary lag may be found for the non-self-regulating process. The area under the curve is a function of the integrating time constant and can be used to find it:

$$\int (c - c_0)\, dt = \Delta m_1 \tau_w^2 / \tau_1 \tag{6.4}$$

Any secondary lag reduces the peak height and therefore can be estimated from it:

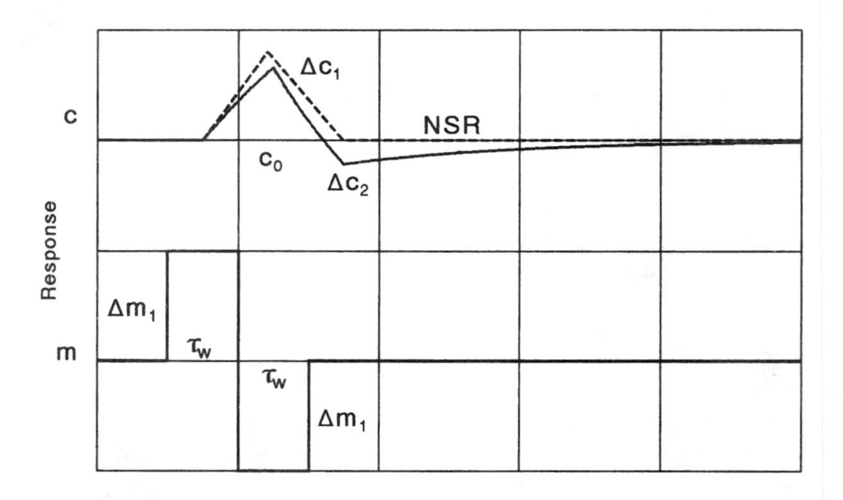

Figure 6.4 A doublet pulse can return even a non-self-regulating process to its original steady state.

$$\tau_2 = \frac{\int (c - c_0)\, dt}{\Delta c_1} - \tau_w \tag{6.5}$$

For self-regulating first-order processes, the two peaks follow exponential relationships:

$$\Delta c_1 = \Delta m_1 K_p (1 - e^{-\tau_w/\tau_1})$$
$$\Delta c_2 = -\Delta m_1 K_p (1 - e^{-\tau_w/\tau_1})^2 \tag{6.6}$$

The overshoot Ω, which is $-\Delta c_2/c_1$, contains only τ_1 as an unknown and can be used to solve for it:

$$\tau_1 = -\frac{\tau_w}{\ln (1 - \Omega)} \tag{6.7}$$

and K_p can be found by combining the last two relationships:

$$K_p = \Delta c_1/\Delta m_1 \Omega \tag{6.8}$$

If the overshoot Ω is 1.0 (or more), these equations cannot be solved, however. This is the case for a process approaching pure deadtime, and the step method must be used as described above.

Finding the secondary time constant is much more difficult for the self-regulating than for the non-self-regulating process. One method is to follow the balanced doublet with an optimal doublet. This doublet is optimal in the sense that the controlled variable is returned to its original steady state in the minimum time rather than following an exponential decay. To do this, its second pulse must be smaller than the first:

$$\Delta m_2 = -\Delta m_1 e^{-\tau_w/\tau_1} \tag{6.9}$$

In response to the optimal doublet, the controlled variable will not cross over its original value and instead will follow a trajectory similar to that of the non-self-regulating process in response to a balanced doublet. (The optimal doublet for a non-self-regulating process is balanced.) This allows Eq. 6.5 to be used to estimate τ_2.

Pseudorandom binary sequences

For those who object to steps and pulses as being too severe for an operating plant to accommodate, a pulse train of a limited amplitude may be preferable. While the response to a small stimulus tends to be smaller still, a repetition of the test can produce correlatable results. For pseudorandom binary sequence testing, a signal-to-noise ratio above 6 is preferred.

The preferred pulse train would be uniform in amplitude, encompassing the normal operating point, but not uniform in time, since that would not test as wide a spectrum as needed, falling into the same trap as frequency-response testing. A random distribution of the pulses can prevent this problem. Yet true randomness for the pulse widths and intervals is not necessarily efficient, for ranges may be covered that are not particularly useful. A pseudorandom sequence appears to be random but is planned. The practice that has developed is to set a time schedule calling for pulses at intervals of one or two deadtimes; when a pulse is due, it is usually left to the operator to execute or not. Since execution is often foregone or delayed, the pattern appears to be random.

Reference 4 describes pseudorandom binary sequence testing of a distillation column. A switching period of 0.5 hour was selected for a given loop. At switch time, the operator looked to see if a move had to be made in order to keep the process within limits. If not, a decision was made whether a move should instead be avoided at that time, such as, for example, when a natural disturbance is developing. Finally, if none of those conditions exist, whether to make a move is determined by a coin flip. Around 20 moves were made for each manipulated variable, thereby consuming about 10 hours for each test. This is more than twice the time required for a complete step test and probably 20 times as long as required for a doublet pulse test. If a significant disturbance occurs during the course of a test, its arrival must be noted and the results following for the next five time constants discarded. Frequent disturbances then multiply the time required to complete a test.

The pseudorandom binary sequence may be sent to a model in parallel with the process, with their responses compared. The model is then adjusted until it responds sufficiently like the process to satisfy the user. This is the method of choice of Professor Richalet and implemented in IDCOM. Interpretation of pseudorandom binary sequence tests in terms of a convolution model is a complex process requiring an "identification package," which is beyond the scope of this treatment.

Closed-Loop Process Identification

In the same paper[1] where they developed their reaction-curve method of process identification, Ziegler and Nichols also presented a closed-loop testing procedure. In applications where the loop cannot be left open, such as, for example, drum-level control on a boiler, this may be the only viable approach. Because these tests are conducted in the closed loop, the information they convey is exclusively dynamic and centered on the period of oscillation. As a result, such tests cannot be used to classify the process into non-self-regulating, lag-dominant, etc., since classification requires information on steady-state properties.

To obtain information relevant to controller tuning, a cycle must be developed. If there is no phase-angle contribution by the controller, then the period of the cycle is determined entirely by the phase lag produced by the process and is therefore an indication of process dynamics alone. The period thus measured is termed the *natural period* of the loop. Tests can be performed using a controller that leads or lags, which will alter the period and thus convey other information that may be useful in light of the phase contribution ultimately desired for the controller. However, the method proposed by Ziegler and Nichols and others avoids this complication by using a controller that has a phase angle of zero. The two simplest controllers having this characteristic are the proportional and the on-off controller.

Proportional cycling

To conduct this test, one needs a pure proportional controller or one that can emulate a proportional controller in producing a phase angle of zero. While a controller having integral action was disqualified as a proportional controller for applications such as averaging-level control, where a fixed bias was needed, this is not a limitation for purposes of testing. Therefore, by setting the derivative time of a PID controller to zero and its integral time as high as possible (or repeats-per-unit-time as low as possible), the proportional mode alone remains active.

Some digital controllers, however, have their *sample interval keyed to the integral setting*. Increasing integral time to maximum thereby increases sample interval to maximum, adding phase lag to the loop. The purpose of proportional cycling is to produce 180 degrees of phase lag in the process, thereby producing a cycle of the natural period. An excessive sample interval obviously violates this purpose. If the controller you intend to use for this test has this limitation, find another or be prepared to apply a phase-angle correction for sampling, as described in Chap. 4.

Move the controller output manually to bring the controlled variable to a desired operating level. Then transfer the controller to automatic with the proportional band at about half of what its optimal setting is likely to be: 100 percent for a flow or liquid-pressure controller; 10 percent for liquid level or temperature on a heated tank or pressure from a reducing valve; 50 percent for temperature on a heat exchanger or vapor or gas pressure on a compressor; 200 percent for a composition or pH controller. It may be necessary to disturb the loop to develop a large enough oscillation to measure—this can be done by moving either the controller's set point or its output while in manual. If there is a deviation, simply changing the proportional band may cause a sufficient disturbance.

The best information will be obtained when the loop is precisely at the stability limit; i.e., the oscillation is perfectly uniform. In this condition, the loop gain $\prod KG = 1.0$ and the loop phase $\Sigma\phi = -180$. The period of the oscillation produced will be the natural period τ_n if a proportional controller is used, in that all the phase lag is concentrated in the process. Furthermore, the product of process steady-state and dynamic gains $K_pG_p = P_u/100$, where P_u is the undamped proportional band. These two pieces of information, τ_n and P_u, are then used to tune the controller.

Several tries may be needed to achieve a perfectly uniform cycle. If either a uniform cycle or the time required to produce one is unacceptable from an operational aspect, a damped cycle may be used, if proper corrections are applied to the period and the proportional band. Even an expanding cycle may be used, since the same corrections fit. However, some rudimentary process knowledge is required for the corrections to be accurate.

For example, if the process has deadtime as the only dynamic element, then the observed period $\tau_o = \tau_n$ and loop gain $\prod KG = \delta$, where δ is the decay ratio expressed as the ratio of two successive peaks to their common valley:

$$\delta = \frac{c_3 - c_2}{c_1 - c_2} \tag{6.10}$$

The relationship is shown graphically for a decay ratio of 0.5 in the upper curve of Fig. 6.5.

Note that there is some confusion in the expression of decay ratio; Ziegler and Nichols have described the upper response in Fig. 6.5 as "quarter-amplitude decay," using c_0 in place of c_2 in Eq. 6.10. The confusion stems from the difference between a proportional cycle and that of the step-load response of a loop having a controller using the integral mode, which produced the lower curve in Fig. 6.5. A proportional cycle is centered around the final value of the controlled variable c_0. The typical step-load response for a well-tuned controller is not—the valley c_2 is usually close to the set point r, while both peaks c_1 and c_3 lie on the same side of it. The lower curve has a decay ratio of 0.25.

If there is a lag or integrator anywhere in the loop, damping extends the observed period τ_o beyond τ_n, and loop gain $\prod KG$ is actually higher than δ. For the case of a process that is non-self-regulating or lag-dominant, $\prod KG = 1 + 0.20(\delta - 1)$ and $\tau_n/\tau_o = 1 + 0.36(\delta - 1)$. For self-reg-

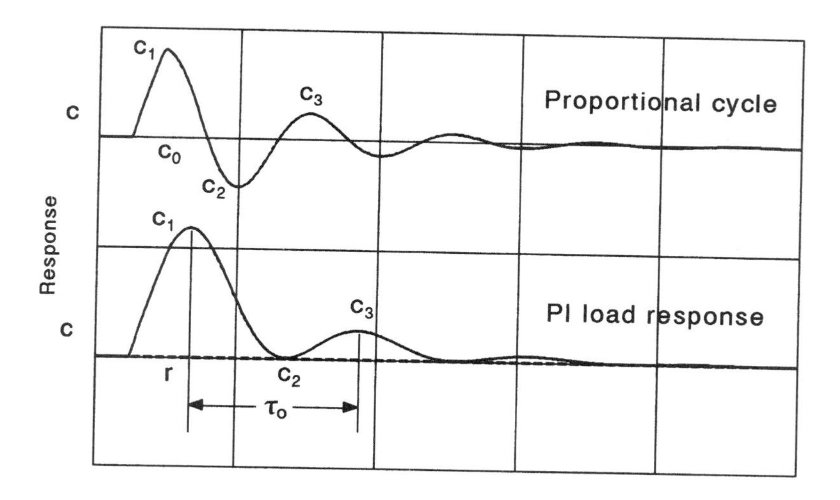

Figure 6.5 The decay ratio is expressed as the ratio of the distance between consecutive peaks and their common valley.

ulating processes falling between the two extremes of lag-dominant and pure-deadtime, some interpolation of the coefficients 0.20 and 0.36 would be required.

Limit cycling

A more refined method of conducting the closed-loop test cycles the manipulated variable between two narrowly defined limits using a proportional gain that is excessive. This procedure has the advantages of limiting the excursion of the manipulated variable and eliminating the search for the undamped proportional band. By means of the high proportional gain, the loop gain within the limits exceeds unity, causing a cycle to expand until the output limits are reached, possibly accomplished in less than one full cycle. A proportional band of 1 to 5 percent would be able to do this for most loops.

Because the oscillation is uniform, the loop gain is in fact 1.00 at the limits, and the observed period should be τ_n. There are some complications, however, in that the limit cycle produced in the manipulated variable is not sinusoidal and that observed in the controlled variable may not be either. In fact, the manipulated variable will tend to follow a square wave, which contains even harmonics as well as the fundamental sinusoid. If the process contained only deadtime, the controlled variable would reproduce the same square wave, and K_pG_p would simply be the ratio of the peak-to-peak amplitude in c to the difference between the controller output limits m_h and m_l.

In the more general case of a process containing multiple lags, the harmonics may be sufficiently filtered that the resulting cycle in the controlled variable is almost sinusoidal. In this event, a correction of $4/\pi$ must be applied to account for the energy delivered by the harmonics. With this correction,

$$K_p G_p = \frac{4}{\pi} \frac{c_1 - c_2}{m_h - m_l} \tag{6.11}$$

If the process were a pure integrator, however, the square wave in the manipulated variable would produce a triangular wave in the controlled variable. To account for the higher ratio of peak height to area for the triangular wave, the correction factor in Eq. 6.11 should then be reduced to $2/\pi$.

If the process load happens to fall exactly between the limits of the manipulated variable, the time spent at each limit will be the same, producing a symmetrical cycle. Only if the cycle is symmetrical will the observed period be the natural period. To ensure this result, the limits should be positioned carefully so that equal time is spent at each. If this is not done, a correction may be required to convert the observed to the natural period, as follows for a non-self-regulating process:

$$\frac{\tau_n}{\tau_o} = \frac{1}{0.5 + 0.25 \dfrac{q - m_l}{m_h - q} + 0.25 \dfrac{m_h - q}{q - m_l}} \tag{6.12}$$

where q is the current load. No correction is required for a process having only deadtime.

A limit cycle also may be created using an on/off controller or relay in place of the proportional controller. While this may be equally effective, it introduces another complication: dead band in the controller. An ideal on/off controller will change states precisely when the deviation crosses zero. Any noise on the controlled variable, however, may cause the output to chatter as the deviation passes through zero. These high-frequency excursions of the manipulated variable could possibly affect the outcome of the test. To prevent chatter, most on/off devices contain some dead band. Like the result of dead band in a valve, changes in controller output proceed via two different paths depending whether the controlled variable is increasing or decreasing. The peak-to-peak amplitude of the controlled variable will tend to increase by the dead band a, and for a non-self-regulating or lag-dominant process, the observed period also will increase by $a\tau_1$.

The results obtained from closed-loop testing, τ_n and either P_u or $K_p G_p$, are to be used in the closed-loop tuning rules that follow.

Open-Loop Tuning Rules

The open-loop tuning rules use process parameters determined from flowsheet information, vessel design data, instrument spans, and valve sizing information, as well as from open-loop tests. They are principally the process steady-state gain, deadtime, and time constants. For multicapacity processes, special rules have been developed based on the number of equal lags, total time response, and whether the lags interact with each other.

To develop these rules, processes were simulated on a personal computer, along with several controllers that were then tuned for minimum IAE using *step load changes applied at the controller output.* (For proportional and PD controllers, IAE is unbounded, so minimum settling time was used instead.) These controller settings are considered to be the most useful for most applications. Other types of disturbances applied elsewhere or using different criteria may produce suboptimal results with these settings. Step set-point response is examined later in this chapter because it is another important consideration, second only to step-load response. However, disturbances of other types or having other points of entry or using criteria other than IAE are not considered, because they are too specialized.

The relationships given below between the optimal controller settings and the process parameters are not based on theoretical considerations such as pole cancellation or model matching. They are empirical rules developed as the results of hundreds of simulations; controller settings were iteratively adjusted until minimum IAE was achieved. The settings thus obtained for a particular class of process and type of controller were plotted against the process parameters, and curves were fitted to connect the data points. Linear, parabolic, hyperbolic, exponential, and logarithmic equations were used to fit the data. Rather than present the data in tables or plot the points, the fitted equations are given in computer code as being the most useful form to present such a large compilation of data.

Equations have been formulated for all self-regulating and non-self-regulating processes. Logic statements are used to separate the functions into groups and to place limits where necessary. The language of choice is QuickBASIC or QBasic; these two languages have the same syntax, but the former is a compiler and the latter an interpreter. (Their ability to structure the code into IF...END IF blocks and CASE blocks is a great improvement over earlier versions of the BASIC language.)

In the following code, all variables are single-precision unless designated as a string by a trailing $. String variables are contained in double quotes ("), and comments follow a single quote ('). T is substituted for the τ of the text because the latter is not allowed in a variable name, and subscripts are simply appended. For example, deadtime τ_d appears in the code as Td; dt is the sample interval. Function EXP(x) is e^x, and LOG(x) is its natural logarithm; ABS(x) is its absolute value; INT(x) takes its integer part; * denotes multiplication, / division, and ^2 squaring. Settings are given in terms of P in percentage proportional band, I as integral time, D as derivative time, and Tc as controller deadtime for the ISA algorithm. Any conversion to other units such as proportional gain or repeats per unit time or to an independent algorithm can be made to the results.

First- and second-order processes

```
OptSettings:

  IF Td THEN T = T1 / Td: TT = T2 / Td 'normalized lags
  SELECT CASE Controller$
    CASE "P" 'Proportional alone
      IF T1 = 0 THEN 'deadtime alone
        P = 200 * Kp
      ELSEIF Process$ = "NSR" THEN 'non-self-regulating
        P = 100 * (Td + .5 * T2) / T1
      ELSE 'self-regulating
        P = 160 * Kp * (1 - EXP(-(Td + .5 * T2) / T1 / 1.6))
      END IF

  CASE "PD" 'Proportional + Derivative
    k2 = 1 - q / 70: IF k2 < .3 THEN k2 = .3 'q is load estimate
    D = k2 * T2 + .5 * Td
    P = 1 + 100 / T: IF Process$ <> "NSR" THEN P = P * Kp

  CASE "I" 'Integral alone; dt is sample interval
    IF T1 = 0 THEN 'deadtime alone
      IF dt < Td THEN I = 1.6 * Kp * Td ELSE I = Kp * Td
    ELSE 'unsuitable for other processes
    END IF

  CASE "PI" 'Proportional + Integral
    IF T1 = 0 THEN 'deadtime alone
      IF dt < Td THEN 'fast sampling
        P = 235 * Kp: I = .5 * Td
      ELSE 'dominant sampling
        P = 125 * Kp: I = .4 * Td
      END IF
    ELSEIF Process$ = "NSR" THEN
      P = 105 * (Td + T2) / T1: I = 4 * (Td + T2)
    ELSE 'self-regulating
      P = Kp * (Td + T2) / T1 * (50 + 55 * (1 - EXP(-T1 / (Td + T2)))
      I = Td * (.5 + 3.5 * (1 - EXP(-T1 / (Td + T2) / 3)))
    END IF

  CASE "PIDi" 'interacting PID
    IF T1 = 0 THEN Controller$ = "PI": GOTO OptSettings
    IF Process$ = "NSR" THEN
      IF TT > .5 THEN k2 = 1.22 - .03 * TT ELSE k2 = 1 + .4 * TT
      P = k2 * 108 / T
      I = 1.57 * Td * (1 + 1.2 * (1 - EXP(-TT)))
```

```
      D = .56 * Td + .75 * T2
    ELSE 'self-regulating
      IF TT > 3 THEN
        P = 40 * Kp / T: I = Td + .2 * T2: D = I
      ELSE
        P = (48 + 57 * (1 - EXP(-1.2 * T))) * Kp / T * (1 + .34 * TT -
          .2 * TT ^ 2)
        I = Td * (1.5 - EXP(-T / 1.5)) * (1 + .9 * (1 - EXP(-TT)))
        D = Td * .56 * (1 - EXP(-1.2 * T)) + .6 * T2
      END IF
    END IF

  CASE "PIDn" 'noninteracting PID
    IF T1 = 0 THEN Controller$ = "PI": GOTO OptSettings
    IF Process$ = "NSR" THEN
      P = 78 / T * (1 + .24 * TT - .14 * TT ^ 2)
      I = 1.9 * Td * (1 + .75 * (1 - EXP(-TT)))
      D = .48 * Td + .7 * T2
    ELSE 'self-regulating
      IF TT > 3 THEN
        P = 30 * Kp / T: I = Td + .2 * T2: D = I
      ELSE
        P = (38 + 40 * (1 - EXP(-1.5 * T)))* Kp / T * (1 + .34 * TT -
          .2 * TT ^ 2)
        I = Td * (.5 + 1.4 * (1 - EXP(-T / 1.5))) * (1 + .48 * (1 -
          EXP(- TT)))
        D = Td * .42 * (1 - EXP(-1.2 * T)) + .6 * T2
      END IF
    END IF

  CASE "PIτ" 'PI with deadtime
    IF TT > 2 THEN TT = 2
    IF Process$ = "NSR" THEN
      P = 100 / T * (.61 + .34 * TT - .12 * TT ^ 2)
      IF TT <= .5 THEN
        I = 1.6 * Td + 2 * T2
      ELSE I = 2.1 * Td + .8 * T2
      END IF
      Tc = Td * (1.6 + 1.94 * TT - .25 * TT ^ 2) 'controller τd
    ELSEIF T <= .1 THEN 'deadtime
      P = 100 * Kp: I = T1 + T2: Tc = Td + T2
    ELSE 'self-regulating
      ln = LOG(Td / T1)
      SELECT CASE T
        CASE IS < 1 'deadtime-dominant
          P = 100 * Kp * (.46 + .36 * ln - .05 * ln ^ 2) * (1 - .22 *
          TT)
        CASE 1 TO 5 'no dominance
          P = 100 * Kp * (.45 + .34 * ln + .08 * ln ^ 2)
        CASE IS > 5 'lag-dominant
          P = 65 * Kp / T * (1 + .15 * TT)
      END SELECT
      I = Td * (.77 - .47 * ln + .05 * ln ^ 2) + .3 * T2
      IF I < .1 * Td THEN I = .1 * Td 'low limit
      Tc = Td * (1.3 - .13 * ln) + 1.5 * T2
    END IF

  CASE "PIDτ" 'interacting PID with deadtime
    IF T1 = 0 THEN Controller$ = "PIτ": GOTO OptSettings
    IF Process$ = "NSR" THEN
```

```
      P = 100 / T * (1.08 - .16 * TT)
      I = .15 * Td + T2 * (.52 - .07 * TT)
      D = 1.5 * Td + .75 * T2
      Tc = 1.06 * Td + T2 * (.84 - .21 * TT)
    ELSE 'self-regulating
      SELECT CASE T
        CASE TT 'identical lags
          IF T < .1 THEN
            P = 100 * Kp: I = T2: D = T1: Tc = Td + T2: GOTO Last
          ELSEIF TT < .5 THEN
            P = 100 * Kp * (1 - .8 * TT)
          ELSE P = 25 * Kp / TT
          END IF
          I = .67 * T2: D = Td * (1 - EXP(-T / .7))
          Tc = Td + .3 * T2
        CASE IS < .5 'deadtime-dominant
          k1 = .133 + .187 * T2 / T1: k2 = .333 + .627 * T2 / T1
          P = 100 * Kp * (1 - k1 * T - k2 * T ^ 2)
          I = .4 * T2: D = T1 + .3 * T2: Tc = Td + .5 * T2
        CASE .5 TO 2 'no dominance
          P = 100 * Kp * (.82 - .2 / T - .3 * TT) / T
          I = Td * (.3 - .35 / T + .1 / T ^ 2) + .55 * T2
          D = Td * (1 - .25 / T)
          Tc = Td + T2 * (.79 -.35 * TT)
        CASE ELSE 'lag-dominant
          P = 100 * Kp * (1.07 - .7 / T - .25 * TT) / T
          I = .15 * Td + .55 * T2
          D = 1.5 * Td * EXP(-Td / T1) + T2 * (.42 - .12 * TT)
          Tc = Td + T2 * (.81 - .19 * TT)
      END SELECT
    END IF

Last:

  IF Tc THEN Tc = dt * ABS(INT(-Tc / dt)) 'round up to integer dt
  IF dt > Td / 20 THEN GOSUB SampleCorr
```

The rounding up is required to avoid noninteger values of deadtime in terms of sample intervals. It is achieved by using the INT function, which rounds down but applies it to a negative number so that its absolute value is rounded up. If the rules for the controllers with deadtime compensation seem unduly complex, remember that these controllers are less robust than the others, requiring more precise tuning to elicit their higher performance and remain stable. Other corrections for sampling are given later under the subroutine SampleCorr.

Tuning rules have not been completely developed for the noninteracting PIDτ_d controller at this writing, but its optimal I and D settings are known to be the same as for the interacting version above. Furthermore, these optimal I and D settings are not subject to change with derivative gain, although the optimal P and τ_d settings may be for processes higher than first-order.

If, after applying these rules, the step-load response of the loop is unsatisfactory, there are several possible reasons. For an analog controller, the tuning parameters perhaps cannot be set very accurately. Another possibility is limited accuracy in identifying the process characteristics. Or the curve fitting may be imperfect in the particular data space representing the process being controlled. Tuning then should continue using closed-loop observations such as overshoot, period, decay, and recovery time. These rules are developed under Fine Tuning in the next section.

Multiple-lag processes

As described in Chap. 2, multiple lags may be linked either in an interacting or noninteracting manner. As the number n of the lags τ increases, multiple non-interacting lags approach the response of deadtime, having a dynamic gain approaching unity and a deadtime approaching $n\tau$. By contrast, multiple interacting lags approach the response of the distributed lag, modeled as a dominant lag plus a smaller secondary lag and equal deadtime. The dynamic gain of the latter approaches about 0.08, as determined by closed-loop testing. Because of this great difference in dynamic gains, the tuning rules for multiple-lag processes as a function of n are quite different for the two types.

The dynamic gain of a process consisting of equal noninteracting lags at its natural period can be calculated quite rigorously. If the process is oscillating uniformly at its natural period τ_n, the total phase lag will be precisely 180 degrees, shared equally by all the lags. Each lag will then produce a phase angle of

$$\phi = -\tan^{-1}(2\pi\tau/\tau_n) = -180/n \tag{6.13}$$

which allows a solution for τ_n:

$$\tau_n = \frac{2\pi\tau}{\tan(180/n)} \tag{6.14}$$

The dynamic gain of each lag is the cosine of its phase angle; the product of all the gains is therefore

$$G_p = \left(\cos\frac{180}{n}\right)^n \tag{6.15}$$

With increasing values of n, ϕ becomes small enough that its value in radians, π/n, approaches tan ϕ; substitution into Eq. 6.14 then gives the approximation $\tau_n \approx 2n\tau$. Convergence of Eq. 6.15 to 1.00 takes a much higher value of n. For example, at $n = 20$, $\tau_n = 1.98n\tau$ and $G_p = 0.781$, while for $n = 100$, $\tau_n = 2.00n\tau$ and $G_p = 0.952$. As a consequence, the tuning rules for equal noninteracting lags are developed as functions of $n\tau$ for the controller time settings and G_p as calculated by Eq. 6.15 for setting the proportional band.

The simplest tuning rules for this process, interestingly, are for the PIτ_d controller. The reason is that the optimal performance for this controller on the basis of minimum IAE is always consistent with zero phase lag. As a model-based controller, its best performance is also achieved at a dynamic loop gain of 1.00. As a result of this combination of properties, its optimal proportional band is simply $P = 100K_pG_p$. However, other complications arise, which are best handled using code as before. T represents time constant τ, and angles are in radians.

```
NonintLags:

    Gp = COS(pi / n) ^ n: nT = n * T: ln = LOG(n)
  SELECT CASE Controller$
    CASE "PI"
      SELECT CASE n
        CASE IS < 20: P = (390 - 63 * ln) * Kp * Gp
        CASE ELSE: P = (140 + 19 * ln) * Kp * Gp
      END SELECT
      SELECT CASE n
        CASE 2: I = .62 * nT
        CASE 3: I = 1.1 * nT
        CASE 4 TO 50: I = .8 * nT
        CASE IS > 50: I = (1.18 - .125 * ln) * nT
      END SELECT

    CASE "PIDi"
      SELECT CASE n
        CASE IS < 6: P = (131 * ln - 24) * Kp * Gp
        CASE ELSE: P = 270 * Kp * Gp
      END SELECT
      SELECT CASE n
        CASE IS < 4: I = (.79 * ln -.45) * nT
        CASE ELSE: I = .5 * nT
      END SELECT
      SELECT CASE n
        CASE IS < 4: D = (.3 * ln -.1) * nT
        CASE 4 TO 10: D = .3 * nT
        CASE 11 TO 30: D = (.53 - .1 * ln) * nT
        CASE IS > 30: D = .2 * nT
      END SELECT

    CASE "PIDn"
      SELECT CASE n
        CASE IS < 4: P = (180 * ln - 133) * Kp * Gp
        CASE 4 TO 30: P = 180 * Kp * Gp
        CASE IS > 30: P = (39 + 42 * ln) * Kp * Gp
      END SELECT
      SELECT CASE n
        CASE IS < 6: I = (.65 * ln - .38) * nT
        CASE 6 TO 25: I = (1.6 - .4 * ln) * nT
        CASE IS > 25: I = (.2 * ln - .44) * nT
      END SELECT
      SELECT CASE n
        CASE IS < 6: D = (.1 + .11 * ln) * nT
        CASE 6 TO 20: D = .27 * nT
        CASE IS > 20: D = (.46 - .062 * ln) * nT
      END SELECT
```

```
    CASE "PIτ"
      P = 100 * Kp * Gp
      I = (.49 - .077 * ln) * nT
      IF n > 4 THEN Tc = .9 * nT ELSE Tc = .8 * nT

    CASE "PIDτ"
      P = 105 * Kp * Gp: IF n < 10 THEN P = P * (.8 * ln - .7)
      IF n < 10 THEN I = .2 * nT ELSE I = .1 * nT
      SELECT CASE n
        CASE IS < 6: D = (.25 * ln - .1) * nT
        CASE ELSE: D = (.48 - .09 * ln) * nT
      END SELECT
      Tc = .9 * nT: IF n < 8 THEN Tc = Tc * (.8 * ln - .7)

END SELECT
Td = .18 * ln 'equivalent τd for sampling correction
GOTO Last
```

For the case of interacting lags, there is essentially no change in behavior beyond n of 20. Furthermore, in most processes containing interacting lags, they are many as in distillation columns or distributed as in packed columns and heat exchangers. Therefore, open-loop tuning rules for these processes are given only for the case where $n >$ 20, essentially the distributed case. The time factor, which was $n\tau$ for the noninteracting case, is replaced by the equivalent $\Sigma\tau$, the time required for the controlled variable to reach 63.2 percent of its complete step response; it is represented in the code by SumT.

```
InteractingLags:

SELECT CASE Controller$
  CASE "PI"
    P = 20 * Kp: I = .5 * SumT

CASE "PIDi"
    P = 15 * Kp: I = .25 * SumT: D = .1 * SumT

CASE "PIDn"
    P = 10 * Kp: I = .3 * SumT: D = .09 * SumT

CASE "PIτ"
    P = 9 * Kp: I = .215 * SumT: Tc = .31 * SumT

CASE "PIDτ"
    P = 6.6 * Kp: I = .067 * SumT: D = .157 * SumT: Tc = .16 * SumT

END SELECT
Td = .1 * SumT 'equivalent τd for sampling correction
GOTO Last
```

Adjustments for sampling and filtering

As described in Chap. 4, sampling adds phase lag to a loop in a manner similar to deadtime, and some adjustment will be required for its presence. All the tuning rules above were produced using a sample interval of $\tau_d/20$ or less. In the presence of slower sampling, the following correction will be required to achieve minimum IAE in response to step-load changes introduced at the controller output:

SampleCorr:

```
Cdt = 1 + .57 * dt / Td
IF Cdt < 1.7 THEN P = P * Cdt ELSE P = 1.7 * P
IF Tc THEN 'deadtime controller
  I = I + 2 * dt: D = D - dt
ELSE
  IF Process$ = "NSR" THEN I = Cdt * I * .9 ELSE I = I * 1.1
  D = D * Cdt
END IF
RETURN
```

These are rather simplified approximations and may not be ideal for all combinations of sample interval and process deadtime, particularly when the two are nearly equal.

Corrections to the controller settings also will be required when filtering is added to reduce noise in the controller output or derivative spikes caused by step inputs as from a chromatographic analyzer. The latter case was demonstrated in Fig. 4.5. The simplest way to allow for the presence of the filter is to add its time constant to the process deadtime before calculating the controller settings. The gain of the filter must be low at the frequency of the noise or spikes to be effective. However, because its time constant must be kept small relative to the other lags in the loop, its principal contribution at the period of oscillation is that of adding phase lag, just like deadtime.

Adjustment for load lag

The preceding rules provide tuning that will drive the IAE to minimum following a step change in load applied at the controller output. The assumption is that the load enters the process through the same time constant as does the manipulated variable. This is the most probable situation but certainly not the only one. When controlling temperature or composition in a distillation column, one dominant time constant governs that controlled variable regardless of the source of the input variable—feed rate, reflux, heat input, etc., although deadtimes and secondary lags may differ. The preceding assumption then is valid for this process. However, in a heat exchanger or boiler, the load (fluid to be heated) and the heating medium enter the process on opposite sides of the heat-transfer surface, and those sides may have different heat capacities and therefore different time constants. The shell of a heat exchanger has more heat capacity than the tube bundle, causing the temperature at the bundle outlet to respond more slowly to steam flow entering the shell than to liquid flow entering the tubes.

The peak deviation and integrated error IE_b for best practical load responses formulated in Chap. 2 used a combination of deadtime in the manipulated path and lag time in the load path. If IE_b varies with the load time constant τ_q, it should not be surprising that the IE produced by the controller following a load change also should vary with it. However, the IE produced by the controller is a function of its tuning, which means that the tuning should be changed as a function of the load time constant. Tests conducted on a self-regulating process under PID control showed this to be the case. Interestingly, no adjustments were required in proportional or derivative settings, only in the integral time, which varied inversely with the load lag.

The open-loop tuning rules above may be applied, calculating integral time I using T1 to represent the lag τ_m in the manipulated path. Then that integral time should be corrected as a function of load lag Tq relative to manipulated lag Tm with the following code:

```
TqCorr:

IF Tq <> Tm THEN 'load lag correction required
  IF Tq = 0 THEN
    IF Tm > Td THEN I = Tm
  ELSE
    I = I * (1 + .4 * (Tm / Tq - 1))
    IF I > Tm THEN I = Tm 'upper limit
  END IF
END IF
```

The first line identifies the need for correction. The second avoids a "divide-by-zero" error in the correction equation and applies an upper limit to integral time at the time constant in the manipulated path. This upper limit is cited later in the chapter as being effective for set-point response, in that the set point has no lag in its path unless a filter is intentionally placed there.

Decreasing the load time constant increases IE_b, but the PID controller can approach it more closely. Consider, for example, a self-regulating process whose time constant in the manipulated path is twice the deadtime. With an equal load lag, an interacting PID controller gives a performance (IE/IE_b) of 0.56 when tuned for minimum IAE. Reducing the load lag to zero more than doubles IE_b, but controller performance improves to 0.66, and when the load lag is increased by a factor of 4, IE_b is reduced by almost that same factor, but performance falls to 0.26. The controller's IE changes only by ± 40 percent, as indicated by the integral correction factor above, yet IE_b changes over a much wider range with the load lag.

Closed-Loop Tuning Rules

There are two situations where closed-loop tuning rules are applied. In the first case, they can be used for estimating the initial settings for the controller if open-loop information is unavailable or incomplete. Lacking any other information on which to base initial settings, the test procedure requires the use of only the proportional mode so that the process response will be obtained free of any phase contribution from the controller. Once the controller has its initial settings, however, it may be expected to contribute phase shift, and the loop also will be damped. Observations of period, overshoot, decay ratio, and recovery time then become functions of both the process and the controller. Closed-loop testing in this mode can then be used to make adjustments to the settings relative to their present values. Thus the information derived from the tests will differ in the two cases, as will the use to which it is put.

Initial settings

The following rules make use of the natural period τ_n and undamped proportional band P_u obtained by cycling the process under proportional control. Alternatively, the process gain product K_pG_p from limit cycling may be used in place of P_u because they are related. However, correction should first be applied for the decay ratio if not unity and for any asymmetry in the cycle, as described earlier in this chapter.

Ziegler and Nichols made the observation that τ_n/τ_d was approximately 4 and gave their tuning rules for P as a function of P_u (although they used gain) and I and D as functions of τ_n. The ratio of 4 identifies the process as being lag-dominant or non-self-regulating; a pure-deadtime process has τ_n/τ_d of 2, and self-regulating processes have intermediate values. This assumes no identification of any secondary lag; in closed-loop testing, there is no way to identify one, although deadtime can still be determined following a disturbance. Deadtime can then be used as a third piece of information to supplement the period and proportional band.

The following tuning rules have been developed for first-order processes. Secondary lags affect both the natural period and the dynamic gain at that period, and therefore, these closed-loop tuning rules tend to compensate for their presence. For example, the optimal value of P/P_u for a process with 20 interacting lags is 1.77, which is the same as for a first-order process having a deadtime-to-lag ratio of about 0.13. The latter also qualifies as a first-order model of the former on the basis of step response. The PIDτ_d controller, however, requires integral time to increase with the secondary lag, and this relationship is not accounted for here.

```
OptSettings2:

    TT = Tn / Td

  SELECT CASE Controller$
    CASE "PI"
      P = Pu * (3.05 - .35 * TT)
      I = Tn * (.87 - .855 * TT + .172 * TT ^ 2)

    CASE "PIDi"
      P = Pu * (3 - .32 * TT)
      I = Tn * (.15 * TT - .05): D = .14 * Tn

    CASE "PIDn"
      SELECT CASE TT
        CASE IS < 2.7: P = Pu * (3.73 - .69 * TT)
        CASE ELSE: P = Pu * (2.62 - .35 * TT)
      END SELECT
      I = Tn * .125 * TT: D = .12 * Tn

    CASE "PIτ"
      P = Pu: I = .21 * TT - .42
      IF TT > 2.7 THEN Tc = .42 * Tn ELSE Tc = Td

    CASE "PIDτ"
      P = Pu * (.7 * TT - .4)
      IF TT > 2.7 THEN I = Tn * (.03 * TT - .08) ELSE I = 0
      SELECT CASE TT
        CASE IS < 3.1: D = Tn * (.243 * TT - .48)
        CASE ELSE: D = Tn * (.132 * TT - .03)
      END SELECT
      Tc = Td

  END SELECT
```

No corrections are required for sampling or filtering if the sample interval and filter time do not change between testing and tuning.

The advantage of the closed-loop rules is that they are much simpler than the open-loop rules, and the information which they require is probably easier to obtain. Such information does not have to be as accurate to be effective either. For example, notice that the derivative settings for the PID controllers are functions of τ_n alone, not of process type, and most of the other relationships are only linear functions of τ_n/τ_d. While they also will be effective for the PIτ_d controller, the test provides too little information to properly tune the PIDτ_d controller—tuning four parameters really requires four pieces of information.

Fine tuning

Whatever method is used to arrive at the initial settings is likely to leave them slightly suboptimal. The tuning task will not be complete until the controller settings produce the true minimum IAE in response to a step change in load. Therefore, the load response needs to be evaluated with the initial settings in place. The simplest way to do this is to step the controller output while it is in the manual mode and quickly transfer to automatic before the controlled variable begins to respond. If the controller is able to make this transfer bumplessly, i.e., without a further disturbance to its output, then the effect will be the same as a step change in load introduced at the controller output.

If a means is available for integrating the deviation, then $IAE/\Delta m$ could be reported for each test, and the settings could be adjusted to minimize its value iteratively. Unfortunately, the means for calculating $IAE/\Delta m$ are usually unavailable, and optimization of two to four controller settings by iteration in real time takes too long anyway. To reduce the effort, certain values of features of the load-response curve such as overshoot, period, decay ratio, etc., have been identified as being coincident with minimum IAE. When the controller is tuned to produce target values for these features, a minimum-IAE response is ensured without having to measure IAE.

An indication of the target values for features of minimum-IAE response curves can be obtained by examining those of the best practical curves presented in Chap. 2. One is reproduced in Fig. 6.6 for a first-order lag-dominant process. Remember that the departure trajectory was determined by the time constant in the load path and that

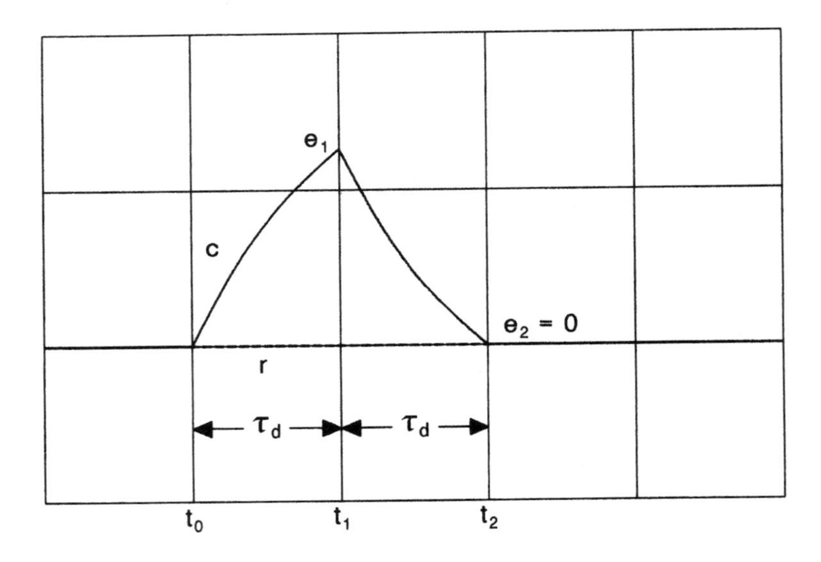

Figure 6.6 The best practical load-response curve has equal departure and recovery durations, along with zero overshoot and decay ratio.

the time interval before the controller could reverse that trajectory is the deadtime in the manipulated path. This time interval, $t_1 - t_0$ in Fig. 6.6, is given the designation of *response time*; the time interval required to recover to set point, $t_c - t_1$, is designated the *recovery time*. In the best practical response curve, they are the same, giving a symmetry S of 1.0:

$$S = \frac{t_c - t_1}{t_1 - t_0} \tag{6.16}$$

where t_0 is the time of departure, t_1 is the time of the first peak, and t_c is the time of crossing the set point. While this response does not appear to be symmetrical, it *is* from the perspective that the peak is located in the center of the curve. With non-self-regulating or higher-order processes or with PI or PID controllers, the optimal curve also will look symmetrical.

The other important features of the best practical curve, its overshoot and decay ratio, are both zero, in that neither a second nor third peak appears. These features are as important as symmetry, but they do not by themselves identify the ideal curve. A long exponential recovery also will have only one peak, but its excessive area is indicated by $S >> 1$.

A more realistic curve appears in Fig. 6.7; it is symmetrical both in time and appearance and reveals a small overshoot and decay ratio. In

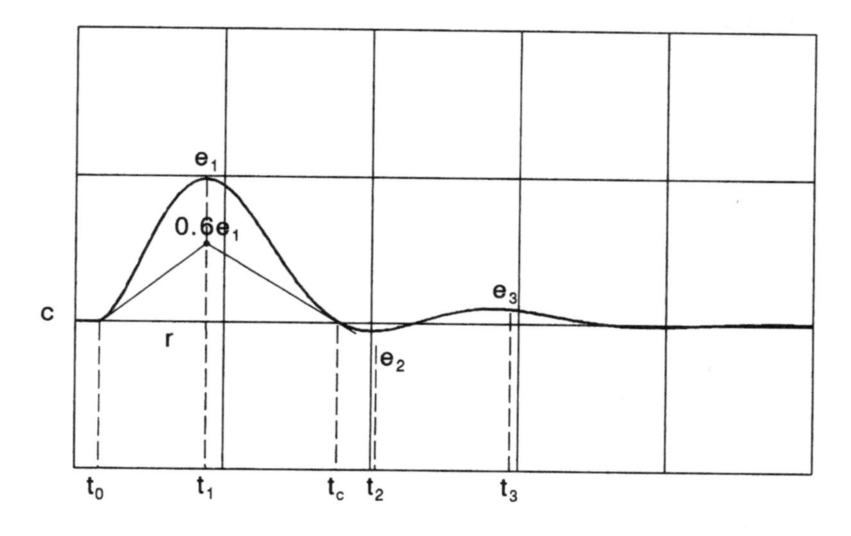

Figure 6.7 A minimum-IAE load-response curve also has similar departure and recovery durations, with overshoot and decay ratio less than 0.1.

a real response curve, it may be difficult to identify t_0 and t_c, the latter being especially a problem for an overdamped curve. To assist in the identification, straight lines are drawn from a point six-tenths of the height of peak 1 tangent to the departure and recovery trajectories; these tangents cross set point at t_0 and t_c. The overshoot $\Omega = -e_2/e_1$, and the decay ratio δ was defined in Eq. 6.10; the period $\tau_o = t_3 - t_1$.

Target values for the features S, Ω, and δ vary with the process parameters and the type of controller used. High-performance controllers can approach the best practical curve with its zero values of Ω and δ closer than low-performance controllers. For the PIDτ_d controller, Ω target increases from zero for a deadtime process to about 0.05 for a lag-dominant process, and δ is essentially unmeasurable. Controllers without derivative action cannot turn corners sharply, and consequently produce more overshoot and a higher decay ratio. For the PI controller, Ω target is about 0.1 for all processes, and δ target is about 0.2, whereas for the PID controllers, Ω target is about 0.05 and δ is unmeasurable. For all controllers, the S target increases from 1.0 to about 1.2 between deadtime- and lag-dominant processes.

With the heavy damping that is characteristic of controllers having derivative action, the period of the well-tuned loop will be difficult to identify, but it is nonetheless a useful feature as a diagnostic of mistuning. The phase lead in an interacting PID controller should cause τ_o/τ_n to decrease from 1.00 for a deadtime-dominant process to about 0.78 for a lag-dominant process; a noninteracting controller can produce about a 10 percent lower period. The phase lag in a PI controller will produce an optimal τ_o/τ_n ratio of 1.35 for all processes. For controllers with deadtime compensation, the loop period must be kept within the notch in the gain curve of Fig. 5.10; this requires τ_o to be precisely twice the controller's deadtime.

Fine tuning based on observed closed-load response requires its own set of tuning rules that relate the required changes in the controller settings to deviations from the target values of the response features. Unfortunately, a change in any controller setting usually affects more than one feature. With this level of interaction, generally applicable rules are difficult to formulate. The problem appears even with the PI controller shown in Fig. 6.8 regulating a lag-dominant process. The upper set of response curves was produced using three different proportional settings, the center curve being optimal. Note that increasing the proportional band reduces Ω and increases S at the same time. In the lower set of curves, increasing integral time also reduces Ω and raises S, but differently. The most noticeable difference is that increasing I leaves e_3 in about the same place both in time and amplitude, while increasing P moves it both down and to the right. For this process, increasing P raises τ_o and lowers δ as well as

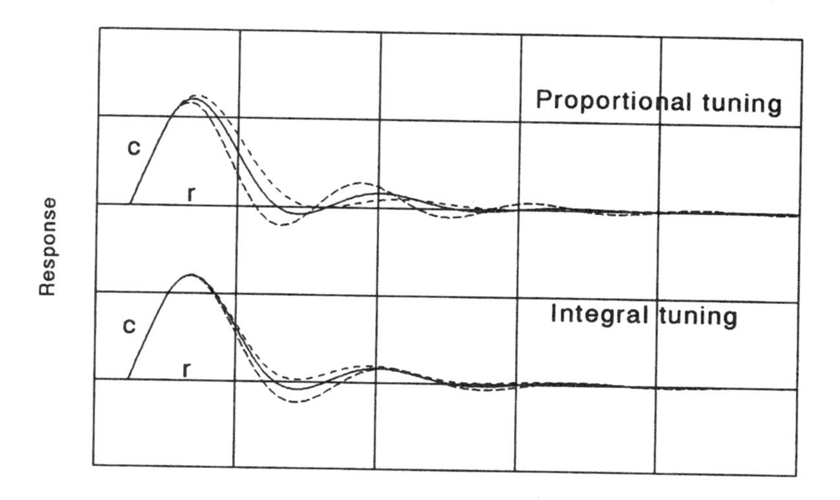

Figure 6.8 Load-response curves for a PI loop with three proportional settings 15 percent apart (*above*) and three integral settings also 15 percent apart (*below*).

Ω, whereas increasing *I* lowers Ω more than anything else. This is not the case for every process, however.

For a deadtime process, changing *P* has no effect on period at all but determines both Ω and δ in a very predictable way. For a deadtime controller, the optimal value of *P* can be determined from its present value using the formula $P_{opt} = P(1 + \Omega)$; the same rule applies to the integral setting for sampled-integral control of deadtime. Also in a controller containing deadtime, a period exceeding twice the deadtime requires an increase in the deadtime setting, and if the period is too short, the setting must be decreased to force the period and the notch in Fig. 5.10 to coincide. If the period is within the notch but the decay ratio is excessive, the preceding proportional-band correction is effective.

Adjusting derivative time using closed-loop response features is as confusing as correcting the integral setting, because around optimal conditions, the two adjustments have similar effects on Ω. This is especially true of an interacting controller, where changing either term also changes the effective value of the other, as well as the proportional gain. Load-response curves for an interacting PID loop appear in Fig. 6.9 using three different derivative settings, the center curve being optimal. They differ from the results shown in Fig. 6.8, where lower *P* and *I* settings increased both Ω and δ, in that a lower value of *D* increases Ω while reducing δ. Note, however, that the

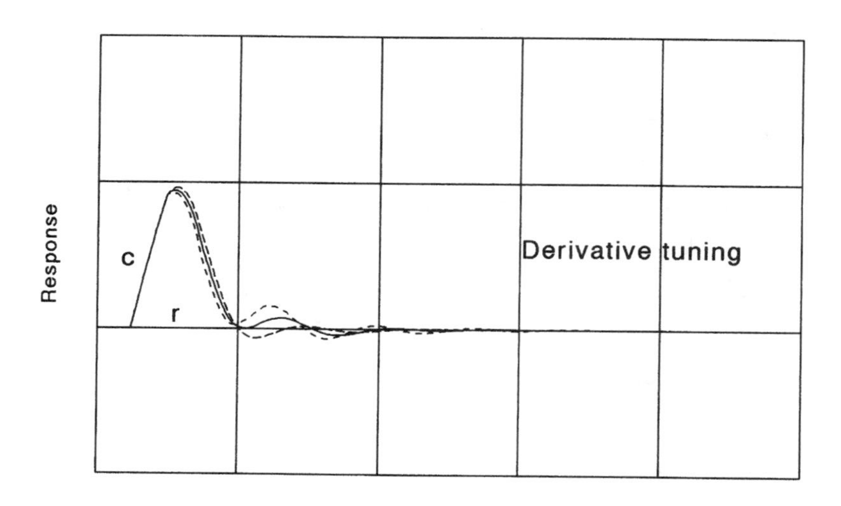

Figure 6.9 Load-response curves for the same process as in Fig. 6.8 under interacting PID control using three derivative settings 15 percent apart.

return trajectories in Fig. 6.9 all cross at about the same place near set point, which should give similar values of *S*, which is not the case for the curves in Fig. 6.8.

One additional relationship should be borne in mind when fine tuning derivative time: It has no impact on integrated error. Therefore, all three curves in Fig. 6.9 have the same IE value. This is not the case for the curves in Fig. 6.8, since IE varies directly with both *P* and *I*. Thus, to reduce the IE in a response curve, a reduction in *P* or *I*, or controller sample interval or deadtime, is necessary. Adding derivative action to a controller can reduce IE only by allowing a subsequent decrease in the integral time.

Tuning for Set-Point Changes

Most of the loops expected to respond frequently to set-point changes are flow loops. They are usually tuned for prompt set-point response with little overshoot, and they are tested by stepping the set point. Load response for these loops is also quite effective with the same settings. Yet the same does *not* hold true for slower loops such as temperature and composition. A temperature controller tuned for optimal set-point response tends to give poor load response, and vice versa. This is a common failing in much of the academic literature—tuning rules based on set-point response produce slow recovery from load changes. There is a reason why flow loops do not have this problem yet others do, and the reason has to do with integrated error.

The integrated-error problem

The controlled variable is delayed in responding to a set-point change by the deadtime and lags in the loop. As a result, there is a net integrated error between the set point and controlled variable, even with the best controller; formulas for estimating IE_b are given in Chap. 2. If the controller has integral action, it will be integrating while the controlled variable is approaching the new set point. When passing from one steady state to another, the controller output is moved an amount Δm proportional to the integral of the deviation, according to Eq. 3.20. The change required in the controller output to respond to the set point is $\Delta r/K_p$. Inserting this requirement into Eq. 3.20 gives

$$IE_r = \frac{\Delta r}{K_p} \frac{PI}{100} \tag{6.17}$$

If this IE_r required to move the controller output matches up well against IE_b estimated for the process, then a favorable set-point response will result. If IE_r is a smaller number, the controlled variable must overshoot the set point, because the controller will have integrated more error during the approach to set point than needed to reposition m the required amount. The area beyond the new set point is $IE_r - IE_b$; it can be made into a dimensionless area overshoot Ω_A by dividing by IE_b:

$$\Omega_A = 1 - IE_r/IE_b \tag{6.18}$$

If IE_r is higher than IE_b, then approach to the new set point would be slower than possible, and overshoot could be avoided.

Since IE_r depends on P and I settings, some retuning of the controller can be done to improve its set-point response. Suppose, for example, that the IE_r estimated using the optimal settings for load response is significantly lower than IE_b. Then, either P or I or both could be increased to produce a better match and thereby minimize the set-point overshoot. This is actually the result whenever controllers are tuned by introducing set-point changes. This gives an integral time that is far too slow for effective load rejection, however. Since IE_b is proportional to τ_1 and IE_r is proportional to I, a popular practice, especially among academics, is to set $I = \tau_1$, called *pole cancellation*. On lag-dominant processes, it always results in slow recovery from load upsets.

The mismatch between IE_b and IE_r is determined principally by the ratio $P/100K_p$. This ratio appeared in all the tuning rules given for load response (except those applied to non-self-regulating processes, where there is no K_p). For a deadtime process controlled by a deadtime controller, this ratio is 1.0; hence there is perfect set-point

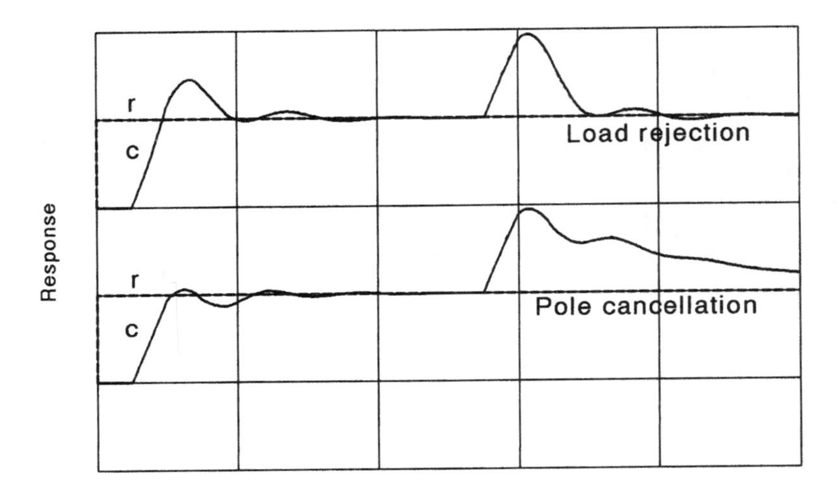

Figure 6.10 Pole cancellation is used to attain set-point response to the detriment of load response (*below*), in contrast to tuning for load rejection (*above*).

response. Moreover, this ratio is typically around 1 for a flow loop. As a process contains more lag, however, the ratio approaches τ_d/τ_1, with the result that Ω_A approaches $1 - \tau_d/\tau_1$.

The pole-cancellation method is illustrated in Fig. 6.10 for a process having a τ_d/τ_1 ratio of 0.2. With the controller tuned for minimum IAE in response to a step change in load, overshoot to a set-point change is almost 0.5 (upper trace). When the integral time is then increased from $1.4\tau_d$ to τ_1 to cancel the dominant pole, set-point overshoot is minimized (lower trace), but recovery from the following load step is exponential at time constant τ_1. The IE for the load-response curve has increased by 5/1.4 or a factor of 3.6, because I has been increased by that factor.

The set-point response problem is at its worst in a non-self-regulating process; lacking any steady-state gain, no change in m is needed to follow the set point, so IE_r is zero. Area overshoot is therefore 100 percent; whether the peak height relative to the set-point step shows this much overshoot is a function of the way the controller settings spread the area out. A PI controller tuned for minimum-IAE load response will overshoot set point by over 0.8; a noninteracting PID controller with its higher gain will overshoot even more. This behavior is unacceptable for many applications.

If the integral action were eliminated from the controller, the set-point overshoot could be avoided; unfortunately, however, any load change would then result in offset, which makes this approach unworkable. The use of a very long integral time does not solve the problem for the non-self-regulating process either; the negative area must still be balanced by a positive area, which although having a smaller peak will be spread out over an extended time. Thus pole cancellation is of no avail here.

The *interacting* PID controller, however, has different integrated errors for set-point and load changes. Consider the interacting controller of Fig. 3.7, having derivative action applied to the controlled variable only. Neglecting the derivative filter, its transfer function is

$$m(s) = \pm\frac{100}{P}\left(1 + \frac{1}{Is}\right)[r(s) - c(s) - c(s)Ds] \tag{6.19}$$

Following multiplication of the terms, replacement of $r - c$ with e, and transformation to the time domain, we have

$$m = \pm\frac{100}{P}\left(e + \frac{1}{I}\int e\,dt - \frac{D\,dc}{dt} - c\,\frac{D}{I}\right) \tag{6.20}$$

Equation 6.20 can be evaluated for the steady states both before the set point is changed and after response is complete, and the difference can be taken to evaluate IE_r between the two states. In both states, e and dc are zero, so

$$\Delta m = \pm\frac{100}{P}\left(\frac{1}{I}\int e\,dt - \Delta r\,\frac{D}{I}\right) \tag{6.21}$$

where Δr has replaced Δc as being equal to it.

Solving for IE_r, we have

$$\mathrm{IE}_r = \Delta r\left(\frac{PI}{100K_p} + D\right) \tag{6.22}$$

For the non-self-regulating process, Δm is zero (or K_p approaches infinity), with the result that IE_r is simply ΔrD. Then, by increasing D, the overshoot may be reduced to any desired level. However, the other settings of the controller also must be changed if the load response is not to be sacrificed. In practice, this is accomplished by swapping I and D and increasing P as needed to keep the proportional gain from changing.

The effectiveness of this method is illustrated in Fig. 6.11 for a first-order non-self-regulating process. The dashed trace is obtained using the optimal load-response PID settings of $108\tau_d/\tau_1$, $1.57\tau_d$, and $0.56\tau_d$, respectively; in the solid trace, they have been changed to $360\tau_d/\tau_1$, $0.6\tau_d$, and $1.9\tau_d$, respectively. Set-point overshoot has been

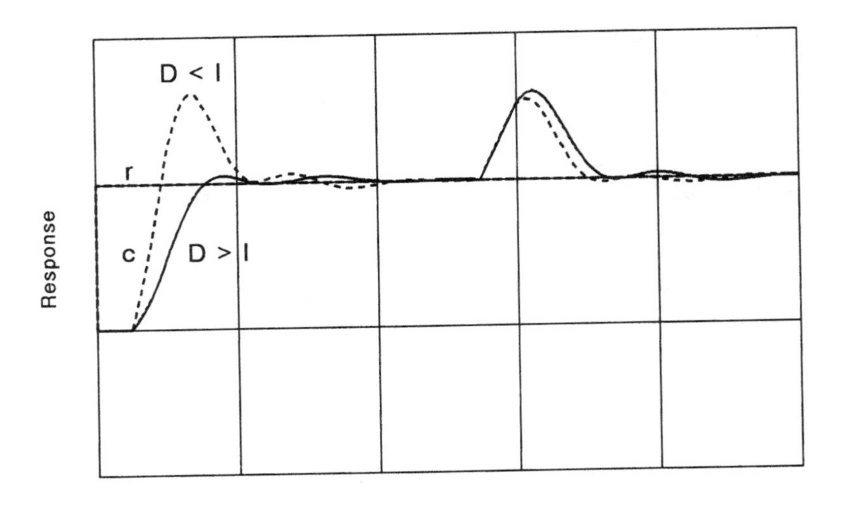

Figure 6.11 Retuning the interacting PID controller to minimize set-point overshoot results in an increase in IAE following a load change.

minimized with only a small loss in the following load response. The *effective* PID values for the two cases are not much different: $79.6\tau_d/\tau_1$, $2.13\tau_d$, and $0.41\tau_d$ for optimal load response versus $86.4\tau_d/\tau_1$, $2.5\tau_d$, and $0.46\tau_d$ for optimal set-point response. Remember that the derivative gain limit K_D is reduced by the D/I ratio, however. For the optimal load tuning, K_D is 8.1, reduced to 2.6 for the optimal set-point tuning. As a result, IAE for a step change in load is increased by 25 percent when using the optimal set-point tuning. This could be alleviated if the derivative filter time were made a function of the effective derivative time rather than its set value.

When tuning the interacting PIDτ_d controller, D is matched to τ_1 for processes that are not lag-dominant, canceling that pole. Even for non-self-regulating processes, it is set much higher than I. As a result, the optimal load settings applied to this controller also produce optimal set-point response. Figure 5.11 shows both responses for the same process controlled in Fig. 6.11. Because it is an interacting controller, IE_r differs from IE for load changes by ΔrD.

Set-point filters

For PI and noninteracting PID controllers, set-point filtering is a method for reconciling set-point and load responses on lag-dominant processes. In essence, the controller is first tuned for optimal load response, and then the filter is tuned for optimal set-point response. The filter usually is implemented as a first-order-lag or lead-lag function.

Honeywell offers two PID algorithms: Type A applies all three PID modes to both r and c, and type B applies only the integral mode to r. Neither type matches the PID controllers described above, where all three modes are applied to c and proportional and integral controls are applied to r. To convert these controllers to the type A algorithm would require applying derivative action to the set point; however, this only exacerbates set-point overshoot. To convert them to the type B algorithm would require inserting in the set-point path a first-order lag having a time constant equal to the integral time. Both algorithms therefore use set-point filtering, either lead D or lag I being the choice. Unfortunately, neither is optimal nor adjustable independent of the PID settings.

Applying a lag τ_r to the set-point increases the IE_r of the controller by $\Delta r \tau_r$. This can be used to provide a better match to IE_b than can the unfiltered controller and can be adjusted for some desirable amount of overshoot. Its principal limitation is that in preventing any proportional action on the set point, the response of the controlled variable is noticeably sluggish. A preferred approach is to use a lead-lag function with dominant lag. The lead term is essentially a fraction G_r of the lag, which also represents the high-frequency gain of the filter. This function changes the IE_r contribution of the filter to $\Delta r \tau_r(1 - G_r)$. Using this filter, proportional response to the set point is not eliminated, but it is reduced by the factor G_r. This allows a faster response than attainable using a lag, with the same protection against overshoot.

Figure 6.12 shows the set-point response obtained for the non-self-regulating process of Fig. 6.11 using a noninteracting PID controller tuned for optimal load response. It has been given a set-point filter equal to its integral time. For the first step, the lead gain is zero, so there is no proportional action on the set point; for the second, the lead gain is 0.2, producing about the same overshoot but 20 percent less IE. More filtering can reduce the overshoot but also retards the response. (Without any filter, the overshoot would be 100 percent.) Using the same filters as in the figure, an interacting PID controller produces about half as much overshoot, with a higher IE due to the contribution of the derivative term.

As described earlier, deadtime-dominant processes respond with about the same overshoot to load and set point and therefore do not require any set-point filter. Between the two extremes of deadtime and non-self-regulating processes, the amount of filtering required varies with τ_d/τ_1. The recommended approach is to fix $\tau_r = I$ and increase G_r from 0.2 to 1.0 in direct proportion to $\tau_d/\Sigma\tau$.

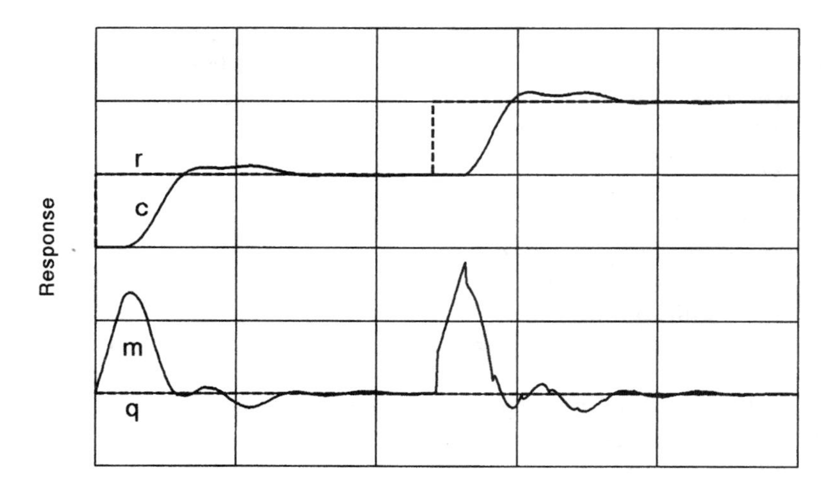

Figure 6.12 The first response is obtained using a set-point lag equal to integral time; the second is obtained using a lead-lag function having a lead gain of 0.2.

Effects of saturation

Unfortunately, filtering the set point is not a complete solution to the overshoot problem. If a large enough step is introduced so that the output of the controller saturates, overshoot will develop even with filtering. To be sure, the proportional bump caused by a set-point change is reduced by the filtering and even eliminated if the lead gain is zero, yet the controller could still saturate if the output is close to its limit when the set point is changed. And a controller that is left in the automatic mode while its output is at a limit will still windup, causing set-point overshoot—filtering is no protection against windup.

There are a number of solutions to the windup problem, some of which are deferred to the last chapter. However, two solutions have in fact already been presented. Using an interacting controller with $D > I$ was found effective in minimizing overshoot, and it remains effective even for a saturated controller. This technique was first described for startup of batch reactors under saturated conditions.[5] A filter puts a lag on the set point, whereas the interacting PID controller has a lead on the controlled variable; the two methods produce similar results as long as the controller does not saturate. When it does, however, the lag on the set point loses its effectiveness, since the controller no longer responds to it; the controlled variable remains active, however, and so its lead is still effective in providing the required IE_r.

A second solution to the saturation problem is the interacting PIDτ_d controller. It protects against overshoot through the same mechanism as the interacting PID controller yet uses the tuning that is optimal for load rejection. This controller offers the same high performance for both load and set-point upsets without regard for saturation.

Tuning for Robustness

In Fig. 5.18, there appeared a comparison of various controllers applied to a particular lag-dominant process based on indices of performance and robustness. The PIDτ_d controller had a higher performance than the PID controllers but a lower robustness, as expected. The performance-robustness tradeoff also appeared between the two PID controllers, along the same diagonal, and yet between the interacting PID and the PI controller, the difference in performance offered no advantage at all in robustness. The lesson to be learned is that there is not always a tradeoff. This was certainly true for the IMC controller, which displayed both the lowest performance and the lowest robustness of all.

Each of the controllers in Fig. 5.18 was tuned for minimum IAE in response to a step-load change (except for the IMC controller, which was matched to the process). In other words, they were tuned for maximum performance. Given that a tradeoff is at least always a possibility, detuning a controller might bring an increase in robustness at the cost of a comparable loss in performance. In practice, the tradeoff is not always there, depending on the process, the controller, and the parameter being detuned. Each case is different and requires examination. Two different processes and their applicable controllers were tested for tradeoff possibilities.

The deadtime process

This case is the easiest to define and follows expectations well. A PI controller optimally tuned has a practical performance of 0.77; its robustness level is 0.97, in that it will tolerate this level of increase in process gain, and its tolerance for deadtime increase is even greater. If the proportional band were then increased by 20 percent of its optimal value, the stability limit would be increased by that same 20 percent. This pushes the robustness index to 1.36, while performance, as measured by IE, falls by 20 percent to 0.65, since IE varies directly with P.

Increasing integral time by 20 percent is not as effective; the stability limit only rises by 6.6 percent, bringing robustness to 1.10. Loss in performance is the same for the change in proportional band, however, because P and I have the same effect on IE.

A PIτ_d controller can approach a performance level of 1.00 on this process, but robustness is limited to a one-sample-interval decrease in process deadtime and no increase—in essence, zero robustness. For example, consider the case where process deadtime is exactly 20 sample intervals. If the controller deadtime is also set for $20\Delta t$, then the loop will be stable while $19 < \tau_d \leq 20$. Increasing the sample interval can improve the robustness, but the controller deadtime setting also must be increased to distribute the sample interval equally on both sides of the nominal process deadtime. In the example case, Δt could be tripled and controller deadtime set to $7\Delta t$, which would be 21 of the original sample intervals. The loop will now be stable for $18 < \tau_d \leq 21$ of the original sample intervals, increasing the robustness level from 0 to 0.05.

The effect of this change on performance is due entirely to uncertainty in response to random load changes. As described in connection with Fig. 4.8, a deviation could develop just before the controller samples, just after, or at anytime in between. The average increase in IE from uncertainty is therefore $\Delta t/2\tau_d$. For the case of the original Δt of $\tau_d/20$, uncertainty already reduced performance to 0.976; tripling its value then reduces performance further to 0.930—a 0.046 loss in performance for a 0.05 improvement in robustness.

Filtering achieves a better tradeoff. Adding a first-order filter of time constant $\tau_d/10$ while leaving Δt at $\tau_d/20$ increases robustness to 0.20, with a loss in performance from 0.976 to 0.899. The filter is therefore twice as productive as the sampler in adding robustness. The other tuning parameters, P and I, have no effect whatsoever on robustness in this loop.

The non-self-regulating process

The non-self-regulating process lies at the other end of the spectrum from the deadtime process and therefore displays different properties of loop robustness and performance. Self-regulating processes will fall in between these two extremes, in the direction depending on whether their lag or deadtime is dominant. Four controllers were tested on a first-order non-self-regulating process: PI, both PIDs, and the interacting PIDτ_d. The effects of detuning the various modes are quite mixed and require individual interpretation.

The PI controller, optimally tuned, has a robustness level of only 0.35 on this process, the stability limit reached by increasing process deadtime; performance is 0.235. Raising the proportional band by 20 percent of its value raises robustness to 0.61, while performance falls to 0.193. Increasing integral time the same 20 percent has the same effect on performance because it has the same effect on IE, but robustness is improved only to 0.45. Therefore, detuning P is the better move to make, even though the stability limit is reached by increasing process deadtime.

The interacting PID controller has exactly the same robustness as the PI controller but a higher performance at 0.571, a fact also pointed out for the lag-dominant loop in Fig. 5.18. (This removes the excuse for not using derivative action on the basis of robustness.) Increasing

P by 20 percent increases robustness to 0.50 while decreasing performance to 0.475. The stability limit is reached first by increasing process deadtime for both settings. Again, increasing I by the same 20 percent is not as effective; robustness rises only to 0.43 for the same performance loss. With this adjustment, however, the stability limit is reached by a smaller decrease in process time constant than an increase in deadtime.

Adjusting D has no effect on performance, as measured by IE, because it has no effect on IE. However, IAE increases as D is moved in either direction from its optimal value because tuning was selected to minimize IAE. Decreasing D by 20 percent has no effect on robustness, although decreasing it further or increasing it causes robustness to *decrease*. On a reduction in D, the stability limit is reached first by an increase in process deadtime, whereas on increasing D, the closer stability limit is for a decrease in process time constant. Decreasing D while increasing I by the same percentage produces essentially the same result as increasing I alone. Therefore, to increase robustness for this controller, raise P and leave the other settings alone.

The noninteracting PID controller has a higher performance on this process at 0.658 and a lower robustness at 0.30. Increasing P by 20 percent raises robustness to 0.52, placing it even higher than the detuned interacting controller, along with a higher performance at 0.546. However, increasing I by 20 percent produces a surprisingly *lower* robustness at 0.28. Reducing D for this controller by 20 percent actually increases robustness to 0.32 with no effect on performance, but raising D by 20 percent plunges robustness to 0.18. Again, the same message comes through: Increasing P is most effective.

For the interacting PIDτ_d controller, performance is 0.769 and robustness 0.10, the closest stability limit being reached on a decrease in process deadtime. Changing either P or I does not move this limit at all, and adjusting D has the same effect as with the interacting PID controller. Filtering is required on a first-order process, with a result similar to that on the deadtime process. Setting the filter at $\tau_d/10$ improves robustness to 0.20 while dropping performance to 0.714. As with the deadtime process, increasing Δt is less effective than filtering. If Δt is tripled from $\tau_d/20$ with no other changes, for example, robustness improves to 0.15 with little effect on performance, but IAE increases due to increasing overshoot and decay ratio. Raising controller deadtime by one sample interval eliminates the overshoot problem but reduces performance to 0.685, and robustness is now 0.125, since decreasing process deadtime represents the closer stability limit.

The presence of a secondary lag in the process reduces performance but improves robustness. For example, if $\tau_2 = \tau_d$, performance of the interacting PIDτ_d controller is 0.490 and robustness 0.28. Filtering now is ineffective, because high frequencies caused by deadtime mismatch are already filtered by τ_2 and *I*. However, robustness can be improved by detuning *P*, although the closest stability limit is reached by changing deadtime; an increase of 20 percent of value improves robustness to 0.40 while dropping performance to 0.403.

While the preceding evidence seems to favor detuning *P* to improve robustness in most situations, there is a caveat. In Eqs. 3.7 and 3.8, the period of a loop comprised of a PI controller and non-self-regulating process was shown to vary directly with the square root of *P* and the damping factor to vary inversely with it. In other words, detuning *P* extensively on non-self-regulating and lag-dominant processes can cause the period and decay ratio actually to increase. Detuning of *P* on these processes should therefore be limited to 20 to 40 percent of value, or else detune both *P* and *I* together to avoid this problem.

Notation

a	Dead-band width
c	Controlled variable
c_n	Size of controlled-variable peak n
d	Derivative operator
dt	Sample interval
D, D	Derivative time
e	Exponential operator 2.718
e_n	Size of deviation peak n
G_p, Gp	Process dynamic gain
G_r	Lead/lag ratio of set-point filter
I, I	Integral time
IAE	Integrated absolute error
IE	Integrated error
IE_b	Best practical IE
IE_r	IE for set-point changes
K_D	Derivative gain limit
K_p, Kp	Process steady-state gain
m	Manipulated variable
m_h	High limit of m
m_l	Low limit of m
n, n	Number of lags
pi	3.1416
P, P	Proportional band

P_u, Pu	Undamped proportional band
q, q	Load
r	Set point
s	LaPlace operator d/dt
S	Symmetry of load-response curve
t	Time
t_c	Time of crossing set point
t_0	Time of departure from set point
t_1	Time of first peak
Tc	Controller deadtime
Δt	Sample interval
δ	Decay ratio
Δ	Difference
Π	Product
Σ	Sum
$\Sigma\tau$, SumT	Sum of the time constants
τ	Time constant
τ_d, Td	Deadtime
τ_m, Tm	Time constant in the manipulated path
τ_n, Tn	Natural period
τ_q, Tq	Time constant in the load path
τ_r	Time constant of set-point filter
τ_w	Pulse width
τ_1, T1	Primary time constant
τ_2, T2	Secondary time constant
ϕ	Phase angle
Ω	Overshoot
Ω_A	Area overshoot

References

1. Ziegler, J. G., and N. B. Nichols, "Optimum Settings for Automatic Controllers," *Trans.ASME,* November 1942, pp. 759–768.
2. Caldwell, W. I., G. H. Coon, and L. M. Zoss, *Frequency Response for Process Control,* McGraw-Hill, New York, 1959, Chap. 7.
3. Shinskey, F. G., *Process Control Systems,* 3d. ed., McGraw-Hill, New York, 1988, pp. 178, 179.
4. Luyben, W. L. (ed.), *Practical Distillation Control,* Van Nostrand Reinhold, New York, 1992, pp. 263, 264.
5. Caldwell, *op.cit.,* Chap. 19.

Chapter

7

Self-Tuning Controllers

Any astute follower of the process-control literature will be aware of the many methods of controller tuning, as well as the many authors of those methods, names well recognized in the field. However, after finishing the preceding chapter, the reader will have noticed the absence of all those names with the exception of Ziegler and Nichols. While inspired by them, all the material presented in Chap. 6 is original, none of it summarizing or reviewing the expertise of others on the subject. Anyone choosing to follow other methods will have to look to other sources for them.

Similarly, there are a multitude of methods for mechanizing tuning under such various headings as *self-tuning, auto tuning,* and *adaptive control.* Like manual-tuning methods, some are more effective than others or more broadly applicable; but in contrast to manual-tuning methods, which are widely reported and discussed in academic circles, commercial self-tuning methods are proprietary, protected by patents or held as trade secrets. Those methods which have not been commercialized are probably not particularly effective, remaining academic curiosities. Even many of the commercial offerings are not particularly effective—most of those observed by me fall into the category of unreliable, if working at all. Therefore, following the approach taken for manual tuning, this chapter will concentrate on the self-tuning methods with which I have experience. Some well-established methods will be missing from this presentation for this reason.

There are two independent purposes for self-tuning control: *commissioning,* i.e., tuning the controller for the first time, and *adaptation,* i.e., retuning to accommodate subsequent variations in process parameters. The former purpose applies to all control loops; the latter, only to nonstationary processes (those whose parameters change significantly), which may represent only 10 percent or so of all controller applications. To commission a controller, it is generally accepted that a known disturbance or series of disturbances may be introduced to identify the process parameters. Those controllers described as having "auto tuning" generally are capable of only the commissioning exercise. In the adaptation mode, intentional introduction of disturbances should be avoided as being counterproductive to control. The real challenge in designing a true "self-tuning" controller lies in interpreting the response of a loop to naturally occurring load changes in terms of required corrections to the controller settings to restore optimal performance.

Pretuning

As described in the preceding chapter, there are two very different approaches to estimating the initial settings for a controller: open-loop testing and closed-loop testing. Both have been automated successfully and are embodied in several control products. They both must be considered "pretuning," however, in that neither produces controller settings accurate enough to eliminate the need for further fine tuning. In effect, they take us from a position of no information to one where initial proportional, integral, and derivative (PID) settings are available, providing at least some measure of control and serving as a basis for further adjustments.

An engineer embarking on the commissioning procedure will be advised to gather as much a priori information on the process as is available in terms of a mathematical model, valve characteristics, residence time, etc., to minimize the amount of testing required. For example, a non-self-regulating process is usually identifiable by its appearance, with most liquid-level loops falling into this category. Also, the behavior of flow loops is so narrowly defined that an experienced engineer will not require any open-loop testing but will introduce initial PI settings and obtain a set-point response immediately. From this response, corrections to proportional and integral settings may be easily made. The task of building a self-tuning flow controller is not particularly difficult.

Unfortunately, other loops are less predictable, and a general-purpose self-tuning controller must be prepared for them all. This is where some auto-tuning controllers are observed to fail. For example, one designed to control temperature cycled continuously when controlling flow. It might be a distinct advantage for a user to identify the type of loop to the controller, at least to set the proper time frame for the expected response. For example, the Foxboro EXACT self-tuning controller[1] has an internal timer that determines how long it will wait following a first peak in a response curve for a second peak to appear before abandoning its search. As the controller leaves the factory, this timer has a default value of 5 minutes, too short for a slow composition loop and too long for a fast flow loop. A user attempting to self-tune a loop at either end of the spectrum using the default timer setting would be disappointed in the results. Accompanying instructions tell the user to follow a "pretune" procedure before attempting self-tuning, which after estimating process deadtime, would set the timer at five times its value. Unfortunately, instructions are usually read after the fact.

Lacking any input as to the type of loop, or even with it, some testing is required. As with manual tuning, both open-loop and closed-loop methods are common. In fact, they are essentially the automation of the manual tests.

Estimating noise level

The open-loop test is initiated on demand from a steady state with the controller in manual. Before introducing any type of test input to the process, however, the controlled variable must first be checked for noise and drift. *Noise* is defined as variations in the controlled variable in a frequency band too high to be reduced by any controller action—essentially variations having a period shorter than two deadtimes. In a digital environment, where self-tuning controllers operate, the sample interval effectively determines the noise bandwidth. Therefore, the first order of business is to specify the sample interval of the controller relative to the expected deadtime of the process. A minimum of 10 samples should represent one deadtime, with anywhere from 20 to 50 giving better results. This requires an estimate of process deadtime, but it need not be very accurate, since the preceding range is wide. Still, it is vital to distinguish between loops having a deadtime of 1 second, 10 seconds, 1 minute, 10 minutes, etc. If the actual deadtime turns out to exceed the expected range, the sample interval can be adjusted and the procedure repeated.

Noise level is probably best represented as the standard deviation σ_c appearing in the controlled variable c over one deadtime. Having selected the sample interval as directed above, 20 samples should be used to make the estimate. Any of the common formulas for obtaining the standard deviation of n equally weighted samples will suffice. If the result is determined only at intervals of n samples, the information is not continuous or current, however.

The result may be updated at each new sample by dropping the oldest as the newest is added. A simpler computation uses the exponentially weighted moving average as described by Eqs. 1.6 and 1.8. It will be recalled from Chap. 1 that this method weighs each sample progressively less with age. The time constant used in this context represented downstream process capacity, but n could be set at 20 samples for use in this application as well.

A parameter called the *noise band* identifies the range over which a variable may be expected to stray within a steady state. For a gaussian noise distribution, the noise band would be $\pm 3\sigma_c$, within which c would appear 99.7 percent of the time. The distribution of "white noise" is uniform or flat, each value having the same probability of occurrence as any other value, up to its amplitude limit. A random-number generator is commonly used to simulate white noise; its noise band is $\pm 1.74\sigma_c$. Uniform distribution is observed in casting one die, since all six numbers have equal probability of turning up; casting two dice produces more-gaussian results, because the probability of the sums differ, with 7 having the highest.

The prominent features of a response curve such as peaks need to be referred to the noise band to determine their significance. For example, suppose the second and third peaks in a load-response curve fall within the noise band. They are then too small to be used to estimate the overshoot and decay ratio, except to note that the overshoot must be less than the noise band divided by the height of the first peak. As a consequence, the controlled variable really needs to exceed the noise band by a factor of 4 or more to provide reliable information on which to base tuning changes.

Because of this relationship between the noise band and the amplitude of significant features, a minimum noise band is best. If the noise level on the controlled variable is zero, however, the noise band should not automatically reduce to zero. It must have a fixed low limit of, for example, 0.2 percent of scale as an arbitrary definition of a significant signal. Because noise level can change with load, set point, and time of day, the noise band should be updated frequently, during every steady state if possible.

The presence of noise complicates efforts to measure rates of change, such as determining the maximum slope of the reaction curve to a step change. While it is relatively easy to draw the best straight line manually through a rising response curve, automating the process using noisy data is quite difficult. Noise also accentuates the peaks of a response curve. Again, it is easy to draw a smooth curve through the mean of the data and quite another to automate the process. The simplest approach is to reduce the observed peak values by σ_c, the most probable error during the short peak transient.

Because the derivative of a signal is especially sensitive to noise, more accurate characterization is achievable using the signal and its integral over time. However, integration is sensitive to drift, which contains very low frequencies, well within the control band. Drift is natural to non-self-regulating processes in the open loop. Therefore, before beginning an open-loop test, a drift check should be made by observing any change in the mean value of the controlled variable between two sets of 20 samples. If the change exceeds the noise band, the test should be delayed until it does not or, in the case of a non-self-regulating process, until the process is rebalanced manually.

Dead band and inverse response

In the preceding chapter, dead band was described as interfering with open-loop test results in removing an indeterminate portion of the test step or pulse. This effect can be overcome by preconditioning the manipulated variable prior to the actual test, thus ignoring the dead band. It is also possible to arrange the test in such a way as to indicate the actual value of the dead band. The question remains whether either of these procedures makes sense from a control perspective.

If the dead band is forced aside by preconditioning the open-loop test, the results may be misleading and the controller tuning unsatisfactory in that the dead band was ignored. On the other hand, whether the actual dead band is 2.9 versus 3.6 percent, for example, is probably unimportant, since the loop will perform poorly in either case. Flow loops can tolerate dead band, even beyond 5 percent; lag-dominant loops cannot—especially level and pH control—and even 0.5 percent is too much. *With the exception of flow loops, if there is enough dead band to interfere with the test, there is enough to interfere with control.*

Excessive dead band will be indicated when the controlled variable fails to return to a steady state following the application of a pulse. Drift will be excessive and difficult to eliminate manually. Effort spent trying to overcome the dead band–induced errors and arrive at compromised controller settings to accommodate the problem are likely to be wasted. Dead band should be corrected at the source, by installing a positioner on the control valve or closing the loop around it with a flow controller. *Dithering* the valve with high-frequency modulation or noise is not nearly as effective.

Inverse response is caused by a negative lead. It tends to drive the controlled variable initially in the direction opposite from what will be its final steady state. The time spent traveling in the wrong direction is essentially lost and affects the loop much as deadtime.[2] There has been no consideration given to this dynamic element thus far, because it is relatively rare in process control except for a small number of well-identified applications. One of these is the response of liquid level in certain types of distillation-column reboilers to the flow of heat. Closing this loop is quite problematic but usually avoidable, since bottom-product flow is normally available to control level. Another is the response of boiler-drum level to the flow of cold feedwater.

The reaction-curve method described for a step input in Fig. 6.1 will accommodate inverse response, interpreting it as deadtime. This turns out to be an acceptable approximation in both open and closed loops, as ref. 2 describes. Inverse response will tend to confuse the results of pulse-test procedures, however, by inserting another peak into the response curve. While it would be possible to devise a pulse-test procedure that would accommodate inverse response as well as the more common dynamic elements, the question remains whether the effort would be worthwhile.

Step or pulse test?

While a step test has the advantage of identifying inverse response, it also has the problem of noise sensitivity in estimating the slope of the reaction curve. This method is used in the Foxboro EXACT controller (1984–1993). To reduce the noise sensitivity, differentiation is avoided, and the maximum slope of the reaction curve is estimated by connecting a series of chords between an inflection point on the curve and successive points below it, selecting the steepest chord.[3] However, the method depends on the location of the inflection point, which does not exist for a non-self-regulating process and so is somewhat contrived and limited in accuracy. Ziegler-Nichols rules are then applied to the results, which only fit load response of lag-dominant loops, giving less confidence to the method.

The Powers 512 controller uses the open-loop step test to determine its PID settings. However, it waits until a new steady state has been reached for completion of the test. This is ideal for deadtime-dominant processes but unrealistic for lag-dominant processes and unworkable for non-self-regulating processes. Ziegler-Nichols open-loop tuning rules are then applied.

Step tests are limited to producing two pieces of information, which is all Ziegler and Nichols expected from them. This is really not enough to tune a controller accurately nor characterize a process satisfactorily. Instead of attempting to differentiate the results of a step test, it is much more efficient to force the derivatives on the process in terms of pulsing the manipulated variable. A single pulse acts as a first derivative of the step, and a doublet pulse acts as the second derivative. The single pulse was shown in the last chapter to produce three parameters, and a doublet pulse could even distinguish a secondary lag. As a result, the technology of open-loop process identification is moving in the direction of using the doublet pulse. It is theoretically sound and relatively insensitive to noise. The need for using an *optimal* doublet can even be avoided if response curves have been modeled carefully to represent a wide range of processes. (This has been done by myself as needed to commission PIDτ_d controllers but is unavailable for publication; U.S. patents pending.)

Automating the doublet pulse is done quite easily. Beginning from a steady state, the manipulated variable is stepped by a desired amount and held. After the controlled variable has changed by a significant amount—for example, four times the noise band—the manipulated variable is stepped an equal amount in the opposite direction and held for exactly the same length of time. In this way, the two pulses are identical and their areas will cancel. The pulse width is automatically keyed to the deadtime.

Identifying peaks is more accurate than calculating slopes. The controlled variable is monitored, and if its present value is more extreme than any past values, its amplitude and time are stored. When the variable retreats from its extreme value by more than the noise band, that extreme value is tagged as a peak, a noise correction is applied as described above, and the search begins for the next peak using the last peak as a starting point. A process approaching pure deadtime will tend to have very flat peaks so that their time of occurrence is indeterminate. However, the identification procedure given for the doublet pulse in Eqs. 6.4 to 6.8 does not depend on the timing of the peaks.

If two pulses are better than one, more pulses should be even better, which is the theory behind the pseudorandom binary series (PRBS). However, the planning of such a test is much more complex. The period of the series as determined by the minimum time between changes must be a function of the process deadtime and lag to produce satisfactory results.[4] PRBS generation has been automated for developing IDCOM models, although the usual practice is to allow operator supervision as described in the last chapter. IDCOM is a multivariable model-predictive control technology used principally to control refinery operations at a relatively high level and therefore consisting of loops having relatively long time constants. However, it is not adaptive nor self-tuning. The process is identified initially in terms of a convolution model; the controller is not tuned but rather is simply detuned in gain to add robustness through filtering.

Relay cycling

An increasingly common method of commissioning a controller is the use of a relay to cycle the manipulated variable in a closed loop. As described in the preceding chapter, this method has the advantage of keeping the process within bounds during the test and does not require the user to make proportional-band adjustments, as does the Ziegler-Nichols closed-loop method. However, a significant source of error in the period observation is asymmetry in the cycle, as described in Eq. 6.12. Hang and Åström[5] describe a calculated method of biasing the controller output to keep the cycle symmetrical, thereby avoiding this error. Placing an integrator between the relay and the manipulated variable also can produce automatic biasing, but this loop is unstable if the process is non-self-regulating. These authors claim that a process nonlinearity also can affect the symmetry, but without changing the period or gain estimates, as the load imbalance does. Nonetheless, correcting the symmetry of a nonlinear loop by biasing does not change the results.

A second source of error was attributed to dead band in the relay, which is required in the presence of any process noise. Corrections also need to be made for the nonsinusoidal waveform of the manipulated and controlled variable, following Eq. 6.11. With all these corrections, this method is not as accurate as proportional cycling. When the results are in, Hang and Åström simply apply Ziegler and Nichols' closed-loop tuning rules, including setting derivative time equal to one-fourth the integral time, a relationship that is definitely not optimal. They do add a lead-lag filter to improve set-point response, however.

Several commercial controllers use the relay auto-tuning method. For example, the Yokogawa UT35 controller cycles the output over its full range, causing an excessively wide cycle in the controlled variable (not recommended except for low-gain processes), with no apparent correction for asymmetry. When tested on a lag-dominant process, it produced a proportional band five times higher than optimal, giving very sluggish closed-loop performance. The Watlow 985 and FGH S900 controllers cycled their outputs over the full range, too, producing an excessive amplitude in the controlled variable and resulting in unstable PID settings. Large cycles such as these are much too sensitive to process nonlinearities to produce accurate results at the control point.

The Omron E5AF auto-tuning controller cycled its output ± 20 percent and produced much more responsive performance, but only with the *fuzzy intensity* set to zero; at its factory setting of 50 percent, auto tuning produced an unstable loop. (*Fuzzy logic* is used in these controllers for the purpose of suppressing set-point overshoot. It is discussed at more length in the next chapter.) In summary, the effectiveness of these commercial auto-tuning functions was far from satisfactory.

Performance-Feedback Adaptation

Most auto-tuning controllers have no ability to improve on the settings determined by the pretune operation. They cannot then correct for errors or false assumptions in the tuning rules and cannot move iteratively toward better performance. Given that the auto tuning does produce a satisfactory result, however, they can still be used adaptively to control a nonstationary process. The operator must first recognize a deteriorating response and request a repeat of the pretuning exercise. To maximize performance and keep it there, however, requires an ability to measure performance and adjust the controller settings relative to any deviations from target values. This capability separates the true self-tuning controllers from those auto-tuning controllers having only a pretune function.

Pattern recognition

As described in the preceding chapter, there is an association between the shape of a response curve and its integrated error, IAE, and whatever other objective criteria might be used to describe it. A minimum-IAE curve has certain characteristics or target features of overshoot, recovery, period, and decay ratio. Ziegler and Nichols focused on quarter-amplitude decay as being a desirable characteristic, but the true relationships between performance and decay are more complex. Precise target values can now be given, although they do change with the type of process being controlled, as well as with the controller.

The optimal features of a response curve happen to form a pattern recognizable to a user. An experienced user can then quickly determine when a controller is performing up to its capability and when it needs retuning, although a well-defined disturbance may be needed for verification. Perhaps because the technology is readily visualized, the automation of *pattern recognition* seems to be the most successful method for the self-tuning of controllers.

The same features we examine visually are detected in the controlled variable, as shown in Fig. 7.1. The prominent features of the load-response curve were noted in Fig. 6.7; a set-point response curve is added here with comparable notation. Observe, however, that in the case of the set-point change, the initial deviation e_1 represents the first peak. Overshoot Ω is defined as

$$\Omega = -e_2/e_1 \tag{7.1}$$

and decay δ is defined as

$$\delta = \frac{e_3 - e_2}{e_1 - e_2} \tag{7.2}$$

The period τ_o is simply the time elapsed between the occurrence of the first and third peaks.

The Foxboro EXACT controller adjusts its PID settings to keep both overshoot and decay at or below targets specified by the user;

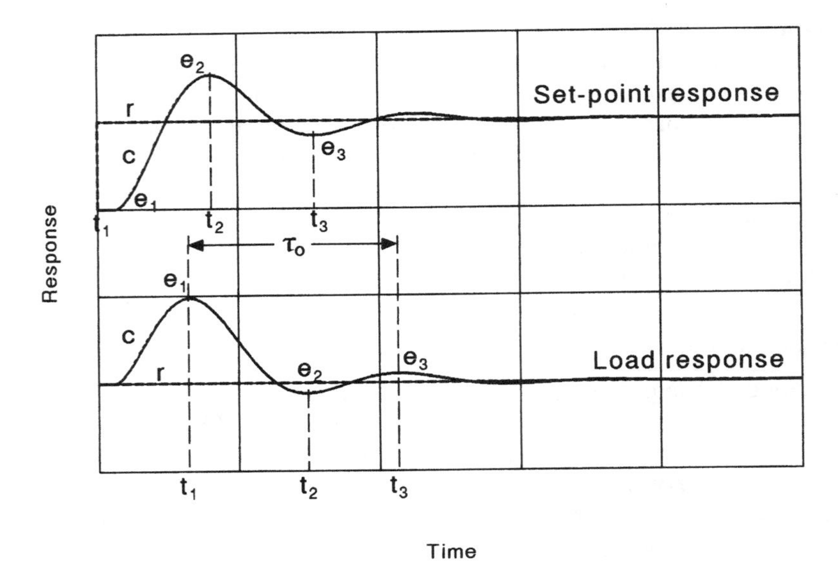

Figure 7.1 For a set-point change, the first peak is the step itself.

that is, it will try to keep one of the two on target and the other below target. In practice, optimal targets for both properties are about 0.1 for PID control if minimum IAE is to be achieved in response to step-load changes. For higher performance, however, as achievable with a PIDτ_d controller, the third peak is usually not measurable, in which case decay is not a useful property. Instead, the symmetry or recovery of the curve as identified in Fig. 6.7 is more meaningful.

If the third peak does not make an appearance, then EXACT is guided only by overshoot information, along with the half-period between the first two peaks. If the second peak does not appear, neither measure of behavior is available, and alternative rules are invoked. The timer mentioned earlier, however, set at $5\tau_d$, must be allowed to expire before the search for the second peak is abandoned. When this happens, it is assumed that the loop is overdamped so that the proportional band is reduced stepwise in an attempt to produce a second peak. Without a second peak, there is simply too little information available to make a judgment on timing, but integral and derivative settings are reduced along with the proportional band.

One failing of EXACT in its original configuration is a tendency to reduce the proportional band of a well-tuned controller when peaks two and three are within the noise band. This raises the decay ratio δ to a visible level above target, requiring a readjustment back to its previous state. Operators annoyed by repetitions of this procedure turn off the self-tuning function after they are satisfied that the controller is properly tuned. This removes the capability for adapting to parameter variations but is satisfactory for stationary processes.

A similar problem arises from the difference between the optimal PID settings for set-point and load response on lag-dominant processes. If EXACT has been tuned to satisfy the overshoot and decay targets for load changes, a subsequent set-point change may produce an overshoot that is well above target. EXACT will retune to reduce that overshoot, but any subsequent load changes will then result in undershoot, requiring retuning in the other direction. Set-point filtering is needed to satisfy the overshoot target for both with the same PID settings.

Convergence toward the targets

Self-tuning forms a feedback loop around the feedback loop being tuned. Therefore, like all feedback loops, it operates iteratively: A deviation from target is observed, an adjustment is made to reduce said deviation, and another observation is made. The self-tuning loop must be much slower than the inner control loop, and it also must be more stable, because it must be able to correct instability within the control loop. To avoid the possibility of overcorrecting the PID settings and possibly causing an undamped or expanding cycle of the tuning loop, EXACT's tuning rules are designed for gradual convergence.

Evidence of this is seen in the succession of adjustments EXACT makes in proceeding from initial to final PID settings. Figure 7.2

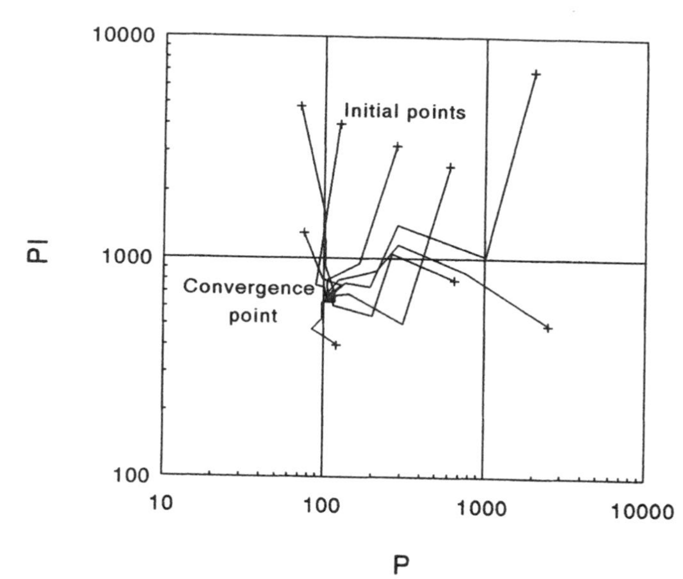

Figure 7.2 These are convergence trajectories for EXACT tuning of a PI controller on a deadtime-dominant process.

shows the results of a series of experiments using EXACT to tune a PI controller on a deadtime-dominant process. Convergence paths are plotted on coordinates of *PI* product against *P*. The product of the *P* and *I* settings will be recognized as proportional to the ratio of integrated error to the size of the load disturbance and hence is an indicator of controller performance. Each + represents an initial set of *P* and *I* values, with subsequent breaks in the trajectories representing the result of each retuning episode. All eventually converge on final values of *P* of 120 and *PI* of 720, indicating the robustness of the algorithm, although several moves are required to reach this point.

While this makes for a stable environment, it does delay reaching the targets and can frustrate an observer who knows that the procedure could be accelerated. Of course, the closer the starting point is to the optimal settings, the fewer moves are required to converge—some of the initial conditions in Fig. 7.2 are a factor of 20 too high in *P*, much farther away than pretuning would produce. Moving *P* alone on these coordinates would produce a diagonal trajectory having a slope of +1. From all the starting points above and to the right of the final position, the initial trajectory is twice as steep, indicating that *I* is reduced by the same factor as *P* in an attempt to produce a second peak.

More accurate modeling of the response of target variables to controller settings can speed convergence without endangering overshoot, but the resulting relationships tend to be nonlinear and restricted as to the range of conditions over which they apply. The corrections tend to take the form of

$$P = P[1 + f(\Omega - \Omega_{tgt})] \tag{7.3}$$

where the new value of *P* is calculated from the old value as a function of the deviation of the property from its target value. Similar corrections apply to *I, D,* and controller deadtime. For some processes such as deadtime-dominant processes, the function may be linear and independent of other parameters; for others, it may be nonlinear and vary with the deadtime-to-time-constant ratio or secondary lag.

For example, overshoot is directly proportional to loop gain when controlling a deadtime process. In this case, Eq. 7.3 can be applied as a straight linear expression with a gain of unity, and convergence requires only one calculation. However, consider a lag-dominant process where the overshoot is excessive. This condition could be due to too low a value of *P, I,* or *D,* as described in Figs. 6.8 and 6.9. It also could be due to too *high* a value of *P,* as is the case in Fig. 7.3. Here the same process is controlled without overshoot using optimal settings, but when the proportional band is too wide, the overshoot increases and the period becomes very long. In observing the slow response curve, there might be a tendency among operators to adjust

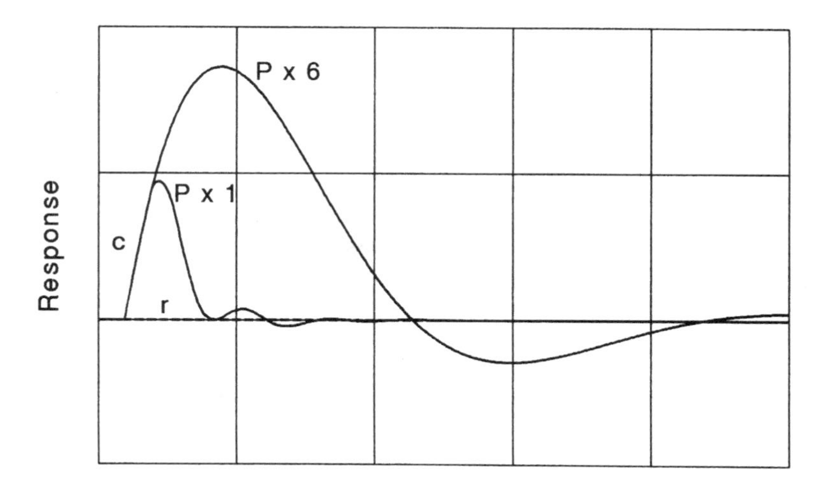

Figure 7.3 Increasing the proportional band above its optimal value can increase the overshoot and period of the controlled variable in a lag-dominant process.

integral and derivative settings as being out of line with the observed period and producing excessive phase lag. However, EXACT will be unable to find the second peak before the timer signals the end of the search, five deadtimes after the first peak. It then assumes overdamped response and reduces the proportional band, as it should.

In observing variations in overshoot or decay ratio and period, there are not enough independent variables to tune *P, I,* and *D* separately. As a result, *I* and *D* are maintained in ratio to each other. This makes sense in that their optimal values given in the last chapter are always related to process deadtime and secondary lag. The ratio D/I changes primarily with process type, approaching 0 for pure deadtime and approaching 1 for a large secondary lag. Once the proper ratio is found for the process type, however, subsequent variations in the process dynamics can develop without changing its type. Consequently, *I* and *D* can be readjusted together as necessary while keeping their ratio constant. The same tends to be true for the PIDτ_d controller, where controller deadtime is included in this combination. Proportional band, however, must always be capable of independent adjustment, because process gain is always capable of variation independent from its dynamic properties.

Once-through processes such as heat exchangers and once-through boilers have deadtime, time constants, and gain all varying inversely with flow.[6] To adapt to changes in flow, all PID settings need to change, so it would be inefficient to move them one at a time.

Sinusoidal disturbances

While performance has been measured principally for step changes in load, real process disturbances are not so well defined. They may include a wide range of frequencies from noise to drift. It is desirable to use a step to test performance from time to time or to initiate retuning, but the retuning procedure also should be effective for naturally occurring disturbances that are not steps.

Sinusoidal disturbances are not uncommon in control loops because sine waves are produced by control loops. In any multivariable system, two controlled variables are likely to respond to the same manipulated variable. If one is enclosed in a loop with the common manipulated variable, that manipulated variable will tend to move in a damped sine wave whenever its loop is upset. The second controlled variable then sees a sinusoidal disturbance imposed by the first loop.

An example of such two-loop interaction is shown in Fig. 7.4. In this particular example, loop 2 does not upset loop 1, but loop 1 can upset loop 2. Initially, a step-load change is introduced into loop 2, and it recovers in an optimal fashion with its PI controller tuned for mini-

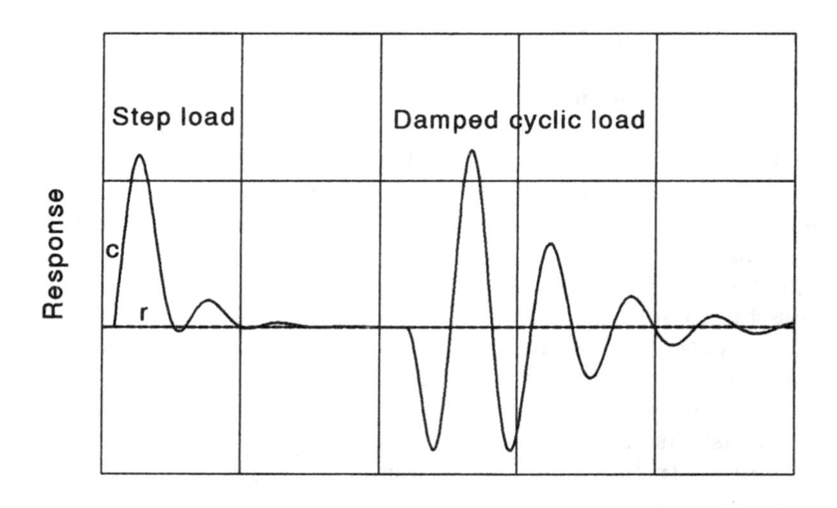

Figure 7.4 This PI loop appears well damped to a step-load change but lightly damped to a sinusoidal upset having its own period.

mum IAE. Later, a step-load change is introduced into loop 1, causing its controller to respond in the same way. However, since its manipulated variable moves in a damped sinusoid, it upsets loop 2, producing a surprising closed-loop response. The second peak in the resulting transient is larger than the first but smaller than the third. This pattern is recognized as abnormal by EXACT, which interprets it as two nonisolated disturbances. Under these circumstances, EXACT is designed to take no action, so it takes none. In practice, the controller has been properly tuned, although this transient does not indicate so.

The reason for the apparently light damping in the last response curve is explained in Fig. 7.5. Here, the sensitivity of a lag-dominant PID loop to load changes is estimated as a function of their period τ_q relative to process deadtime τ_d, on logarithmic coordinates. Disturbances having very short periods are attenuated by the dominant process time constant; those having very long periods are attenuated by the controller's integral mode. In the region around the peak, however, the loop actually amplifies the disturbance, as much as a factor of 2 at the resonant period. The two loops in Fig. 7.4 happen to have identical periods, which is the worst case.

The position of the peak sensitivity in Fig. 7.5 has been moved by retuning. The left and sharper peak corresponds to minimum-IAE settings for an interacting PID controller. Doubling the proportional

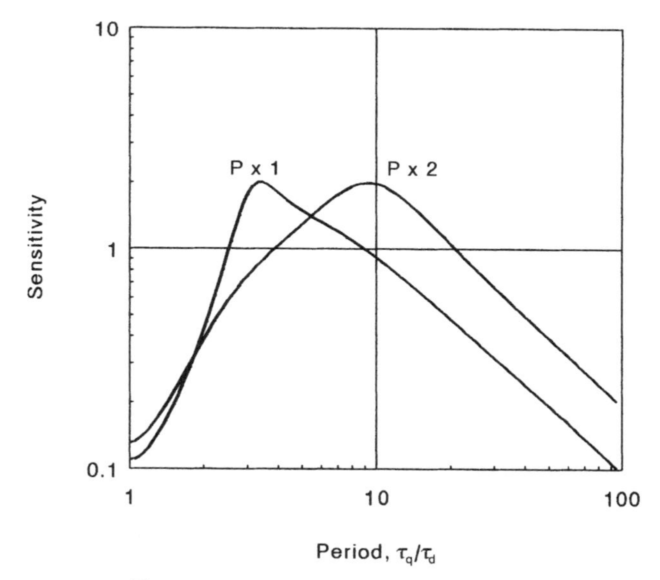

Figure 7.5 The sensitivity of a PID loop to cyclic disturbances exceeds unity around the resonant peak; widening the proportional band moves the peak to a longer period.

band produces the broader peak to the right, raising the entire right-hand portion of the curve. A higher-performance controller will enable the peak to be narrowed, but only within limits. The general nature of the problem is endemic to feedback control. For loops containing identical process dynamics as in Fig. 7.4, some alleviation of the resonance can be obtained by staggering the settings of the two controllers, moving their peaks apart as in Fig. 7.5.

If the periods of the loop and the disturbance differ, the sensitivity is reduced, which is helpful. However, this introduces another potential problem to the tuner, be it a person or a self-tuning function. A cyclic disturbance will tend to force the controlled variable to cycle at the disturbance period, which the tuner could mistake for the loop period. Observing an undamped or lightly damped cycle, the tuner would then widen the proportional band of the controller to lower the decay ratio. Examine what would happen if the disturbance period were above $6\tau_d$ relative to the resonant peak of $3\tau_d$ for the optimally tuned controller in Fig. 7.5. Widening the proportional band would actually *raise* the sensitivity to the disturbance, causing the observed cycle to *increase* in amplitude. This response to a tuning change has scared away more than one engineer.

The timer in EXACT can be used to discriminate between the periods of the loop and the disturbance. If the loop resonates at $3\tau_d$ as with the optimally tuned PID controller, the timer could be safely set to wait for $2\tau_d$ to elapse after the first peak before giving up the search for the second, since the half-period between them would normally be $1.5\tau_d$. Then any disturbance having a period exceeding $4\tau_d$ would not have its second peak detected by EXACT, thereby avoiding the mistake of tuning to the disturbance period.

Monitoring the output

The EXACT controller does not monitor its output except to be aware when limits are encountered, since this can affect the response curves. However, the output of the PIDτ_d controller can contain frequencies that do not appear in the controlled variable. Because of its limited robustness, cycles are very likely to arise from a small mismatch between controller and process deadtimes and will make their appearance in the output first. Consequently, the output of a controller containing deadtime compensation should be monitored continually for cycling, and its decay ratio should be calculated regularly, although it is not normally measurable in the controlled variable.

The target decay ratio for the output varies with process type, from zero for a pure-deadtime process to as high as 0.6 for a lag-dominant process. Tuning rules also vary with process type, as well as the period of the wave. The most difficult challenge in tuning a controller with deadtime compensation is determining the direction of mismatch against process deadtime. The residue from a mismatch has a pulse width equal to the difference between the two deadtimes, and mismatches in either direction are equally possible and virtually indistinguishable. While a filter is essential in removing the highest frequencies, it also reduces performance and should not be considered as a substitute for retuning. Still, a filtered output is far easier to monitor than an unfiltered output.

Effects of Nonlinearities

The most common reason for using a self-tuning controller is the presence of some nonlinearity in the loop. This may not be the best reason or even a good reason, as will be demonstrated. Nonetheless, nonlinearities can interfere with controller tuning and require retuning with changing load.

Output characterization

In most loops, a linear relationship between the controller output and the manipulated variable is desirable to keep loop gain constant. The relationship may be significantly nonlinear if the controller drives a valve, either because the valve itself has a nonlinear characteristic between stem position and delivered flow or because its pressure drop varies with flow.

Some valve types are inherently nonlinear; for example, butterfly and ball valves have a natural logarithmic (equal-percentage) relationship between stem position and port opening. The actual flow delivered also depends on the pressure drop, which can change with flow, thereby altering the effective characteristic of the valve. An equal-percentage characteristic is often chosen to compensate for the pressure-drop variation to produce a more linear behavior. Compensation is always less than perfect, however, and often foregone, so nonlinear manipulation of varying degrees can be expected.

The principal exception is the manipulation of flow by a primary controller through a secondary flow controller in cascade; then the linearity of the *flowmeter* determines the linearity of the primary output, the valve only being seen by the flow controller. A similar result is obtained by manipulating the speed of a metering pump.

It is also possible, however, to have a process that *requires* a nonlinear output characteristic when linear flow is being manipulated. This is the situation in many heat-transfer applications. The sensitivity of a controlled fluid temperature to heat flow varies with the flow of fluid being heated; the same change in steam flow, for example, will raise the temperature of a small flow of fluid more than that of a large flow. Proper compensation for this type of nonlinearity requires manipulating the logarithm of steam flow though an equal-percentage valve. Because these processes are so common, equal-percentage valves are probably used more often than linear valves.

With the existence of all these possibilities, it is not surprising to encounter loops whose gain varies with controller output (and hence with load) in one direction or the other. Because this nonlinearity tends to be consistent and determinable, it should be compensated by a matching nonlinear characteristic. Consider, for example, controlling absolute pressure in a vacuum chamber by injecting air through a butterfly valve. Pressure is linear with flow, but the valve is essentially equal-percentage, exhibiting a gain change as much as 50 to 1 over its flow range. A characterizer needs to be inserted between the controller and the valve, having a curve complementary to the valve characteristic. When properly executed, the result will be flow linear with controller output. Methods for accomplishing the required characterization are presented at the end of the next chapter.

Without proper characterization, loop gain will change with load and possibly with set point as well. Manual tuning must provide stability at the highest gain, which means that low controller performance can be expected at other operating points. Self-tuning controllers have been used to remedy such a situation, but they provide a more costly and less effective fix than nonlinear characterization. A self-tuning controller can tune itself for the current load, perhaps after several iterations, but because the next load will require different tuning, the response of the controller to the next load is likely to be unsatisfactory, followed by another series of iterations. A self-tuning controller is always trying to play "catchup" with the process because it is a feedback loop around another feedback loop.

Input characterization

Probably the greatest challenge represented by a single loop is controlling the pH of industrial wastewater. The control of pH in general is difficult because of the logarithmic relationship between manipulated reagent flow and measured pH. As shown in Fig. 7.6, in the absence of buffering, the titration curve is extremely steep in the neutral range around pH 7, where the set point is usually positioned. If the controller is provided with no nonlinear compensation, then its proportional gain must be low enough to avoid cycling around the set point. With some controllers, it cannot be set sufficiently low, and cycling between pH 5 and 8 or 4 and 10 continues unabated, a fairly common case in industry. Not only is this unacceptable control, but if the controller is manipulating acid and basic reagents in split range, cycling consumes excessive amounts of these reagents.

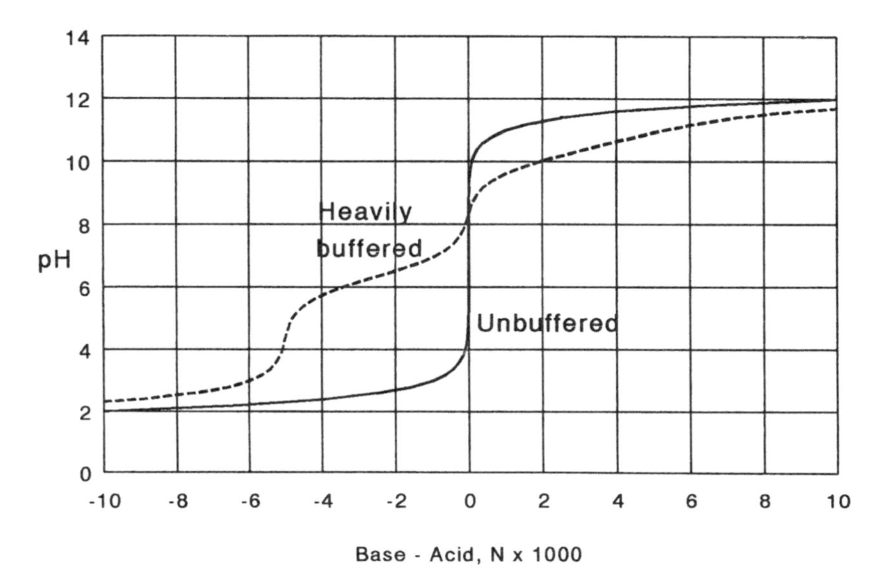

Figure 7.6 The titration curve of industrial wastewater can vary from essentially unbuffered to heavily buffered in a matter of minutes.

If the proportional gain can be set low enough to dampen the cycle, load response tends to be very poor. Following a step change in load, the controlled pH tends to move quickly beyond the specification limit, waiting for the controller to respond. However, its gain is so low that many deadtimes can elapse before the pH is driven back to set point, and because of integration during this interval, a severe overshoot usually results. Allowing the effluent pH to linger above 10 or below 4 for minute after minute like this can cause more damage than limit-cycling can, especially to downstream biologic activity.

A linear self-tuning PID controller can be applied to a pH loop in an effort to cope with its nonlinearity, and it can, but with mixed results. It must raise its proportional band sufficiently to avoid cycling—in the simulation described in Fig. 7.7, P had been raised to 1600 percent before the load change entered. Because of this wide band, the load change drove the pH beyond 10 for several deadtimes. The self-tuning controller simulated here responded to the apparent offset by reducing the proportional band in an effort to reduce the deviation. Eventually, the band was narrow enough to bring the pH back to set point, but this resulted in cycling. The band then had to be returned to its former value of 1600 percent to dampen the cycle. Without nonlinear compensation, each load change requires adaptation, even with a stationary process. Recovery takes place at the adaptation cycle, which is much slower than the cycle of the pH loop itself. This is not an optimal solution to the problem of a nonlinearity in the loop.

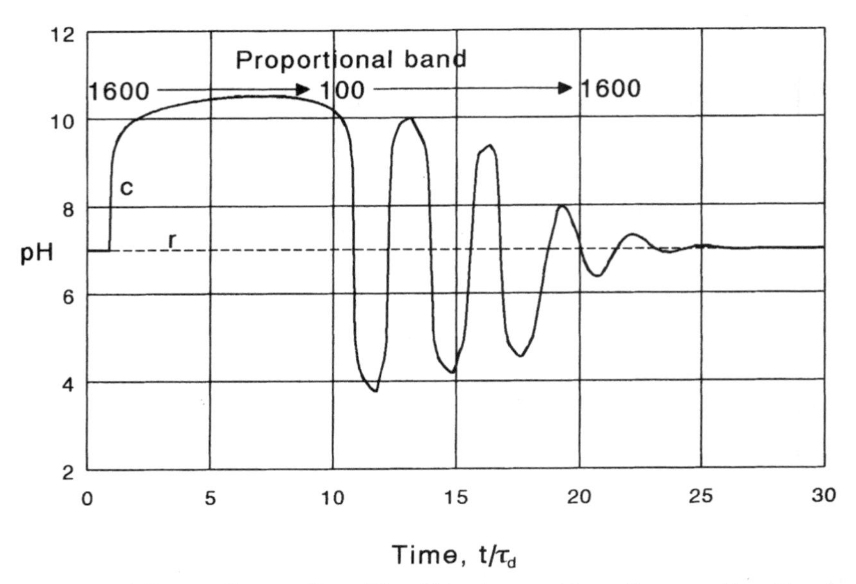

Figure 7.7 A linear pH controller with self-tuning must have its proportional band adapted twice following each load change.

A simple three-line characterizer can be applied to the pH controller, as shown in Fig. 7.8*a*. The center zone has a low gain in the pH range where the gain of the titration curve is maximum; gain outside the center zone is 1; i.e., the normal proportional gain of the controller is applied there. This combination provides the damping required around set point yet forces a more active response to large upsets. It does add two more tuning adjustments: the width of the low-gain zone and the gain within it.

More precise characterization might be justified if tight control were required over a chemical product rather than a waste that simply has to fall within specification limits of typically pH 6 to 9. In this case, the reverse of the process titration curve needs to be inserted in *both* the controlled-variable and set-point paths. In this way, pH measurement and set point continue to be displayed, but the controller is acting on acid-base difference, and compensation is correct for all set points.

There is still another dimension to the pH-control problem, however. *Buffers* in the solution alter the shape of the titration curve dramatically, an example of which is shown in Fig. 7.6. Buffers are very commonplace chemicals and so are present to varying degrees in most solutions. Carbonates common to groundwater are most often encountered, flattening the curve somewhat at pH 6.5 and 10.3. While this makes pH control at 7 easier, the carbonate content may not be constant, and the controller must be tuned for the least buffering observed.

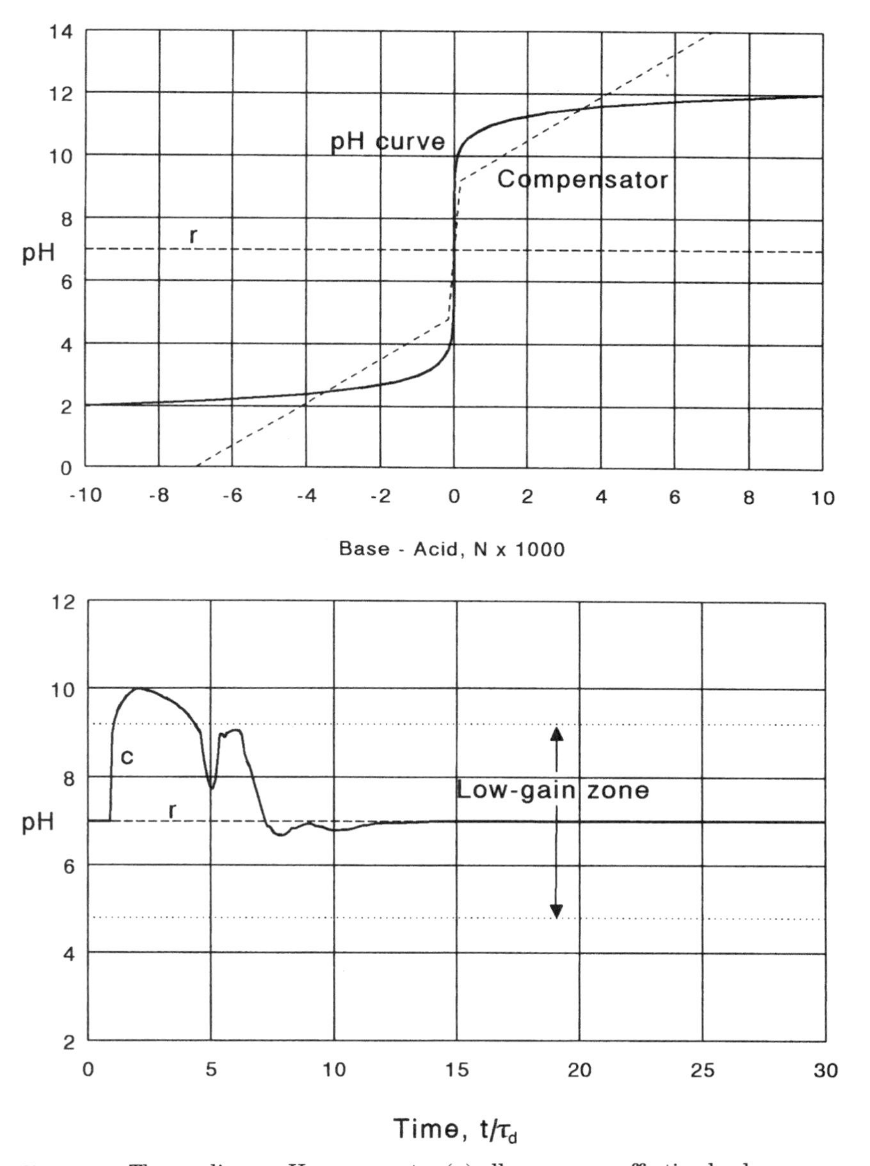

Figure 7.8 The nonlinear pH compensator (*a*) allows a very effective load response (*b*) without self-tuning.

All cleaning agents are buffers, from phosphates and silicates to detergents, ammonia, and amines. Furthermore, most metal ions are also buffers, including iron, aluminum, copper, cadmium, nickel, cobalt, zinc, lead, chromium, etc. As a result, wastewater from metal finishing operations tends to be highly buffered. When ion-exchange beds used to treat boiler feedwater are regenerated, their effluent is also highly buffered, but only at first, the buffering diminishing with time. In other wastewater treatment facilities, buffering is also likely to come and go as individual process vessels are dumped and cleaned. As a result, a fixed nonlinear compensator may not be enough for many applications of pH control—self-tuning also may be required.

Adaptive pH control

Perhaps the first adaptive pH controller was developed by myself in 1973.[7] The nonlinear controller described in Fig. 7.8 was unable to produce consistently acceptable results in the wastewater treatment facility of a particular chemical plant where it was applied because the buffering of the effluent was varying unpredictably. The adjustment having the largest effect on stability seemed to be the width of the characterizer's low-gain zone. It needed to be expanded if an oscillation persisted and contracted if load recovery were slow.

A filter-based method was applied to distinguish between the two situations. High frequencies characteristic of oscillation were rectified positively and integrated to expand the low-gain zone; deviation at low frequencies (all others) were rectified negatively and integrated to close the zone. The time constant of the filter was important, having to be set to pass the loop period; ultrasonic cleaning had to be applied to the pH electrodes to minimize fouling, which would raise the loop period. The time constant of the integrator determined the rate of convergence and had to be set for stable operation.

The first on-line operation of the self-tuner produced "burst cycling," cycles that would expand, then dampen, and then disappear for awhile before returning. This was caused by excessive tightening of the low-gain zone during quiet periods and was corrected by having two different time constants for the integrator, expanding the low-gain zone faster than it could be contracted. This was enough to satisfy the broad specification of effluent pH. This same self-tuning method was used to adjust the proportional band of the controller simulated in Fig. 7.7. Because it cannot be made to adjust the integral and derivative settings of the controller, however, or its own time constants, it is no longer recommended for pH control.

A decade later, Albert and Kurz[8] applied a model-reference approach to the same problem. They compared the rate of reduction of pH deviation to the optimal rate based on observations of the optimally tuned loop. When variations in buffering then changed the rate of recovery, an integrator responded by readjusting the gain of the pH controller. In essence, integral control was applied to the velocity of the pH measurement, and it produced a satisfactory result. This method, however, is also affected by changing dynamics in the loop such as caused by electrode fouling.

The recommended solution for pH control under conditions of variable buffering would use a nonlinear characterizer matched to the most severe (least buffered) titration curve expected. Moderations in the curve can then be accommodated by EXACT or another self-tuning algorithm. EXACT also can accommodate variations in response time due to electrode fouling, which the other adaptive controllers described above cannot, in that they can only change the controller gain. Although the best response will always be obtained by keeping the electrodes clean, some accommodation for deteriorating dynamics is helpful, with an alarm placed on the adapted integral time of the controller to signal when response is no longer acceptable.

Batch processes

Opportunities for adaptive control abound in batch processes, because the batch process tends to change its properties with time, and the equipment may as well. Recipes frequently change, too, which means startup conditions are variable. Adaptation for properties changing with time is no more difficult than for a continuous process, although requirements may be more frequent. The principal problem has to do with unknown initial conditions.

Consider the startup of the batch polymerization reactor shown in Fig. 7.9. The cascaded temperature controllers on the reaction mass and coolant exit can be tuned properly for tight control during the exothermic stage of the operation. However, windup protection also must be provided to avoid overshoot of set point and its effect on product quality and loss of production time. Even this may not be adequate, however, since a properly tuned and protected controller can bring batches to the control point on identical trajectories only if the process parameters are always the same.

One method of windup protection requires an estimate of the load when set point is reached. This can change from batch to batch and definitely changes with set point. The primary thermal time constant of a vessel is identified as

$$\tau_1 = \frac{MC}{UA} \tag{7.4}$$

where M and C are the mass and heat capacity of the reactor and its contents, and U and A are the overall heat-transfer coefficient and area. Fouling can cause U to decrease gradually from batch to batch until the vessel is cleaned. Recipe changes however, also can affect the batch size. If the bottom of the reactor vessel were not jacketed, A would change in

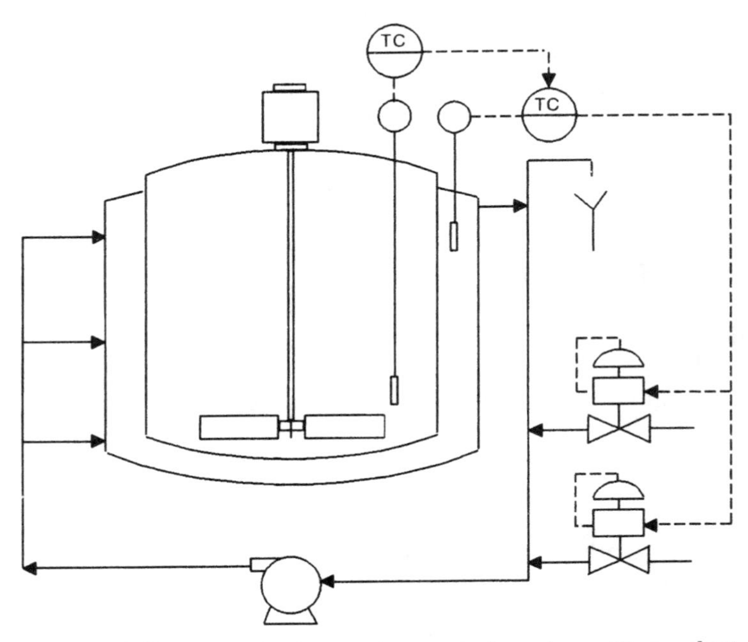

Figure 7.9 Effective reactor temperature control requires setting coolant exit temperature in cascade; heating and cooling valves are manipulated in split range.

proportion to *M*, keeping τ_1 constant with batch size, but this is rarely the case, so τ_1 can be expected to increase with batch size.

I studied the possibility of estimating τ_1 based on the rate of rise of reactor temperature during heating to set point to use as a basis for setting the proportional band of the primary temperature controller. The method proved to be reasonably accurate, providing that no reaction was taking place that would affect the outcome. Pilot-plant tests were then conducted using a vessel accommodating various batch sizes and having a heat-transfer surface divided into two sections. A 4 to 1 variation in τ_1 was produced using different combinations of batch size and surface, and a reasonably accurate estimate of τ_1 followed in each case.

Surprisingly, however, neither temperature controller required retuning as τ_1 was varied, probably due to the configuration of the controllers. Windup protection for the primary controller was provided by connecting the secondary temperature measurement to the integral feedback term of the primary. In effect, this placed the entire secondary loop within the primary integral feedback path. The primary integration rate then is limited to the speed of response of the secondary temperature to the manipulated heat flow. A larger τ_1 causes a slower rise in both temperatures, which the controllers can accommodate without retuning. In summary, this configuration seems to eliminate the need for self-tuning. It does require that the secondary controller have integral action, however, or its offset will be transferred to the primary controller. This configuration is described in more detail in the last chapter of this book.

Batch pH control also may require adaptation if recipe or other changes produce a variety of titration curves. First, a PD controller with zero bias should be selected as described in Chap. 3, along with an equal-percentage valve for logarithmic characterization. Because of the irreversible nature of batch endpoint control, however, a self-tuning controller to correct for an overshoot would act too late. Furthermore, self-tuning methods used elsewhere are not applicable to controllers having no integral action. Therefore, the usual methods of self-tuning do not apply here, and resort must be made to gain-scheduling methods on the basis of recipes or measurable inputs. *Gain scheduling* is simply programming one or more controller settings to vary as a function of some measurable variable such as flow, volume, temperature difference, etc. The functions used tend to be very process-specific and not likely to be found in a general-purpose controller.

Notation

A Heat-transfer area

C Heat capacity

D Derivative time

e Deviation

e_n Size of nth peak

I Integral time

M Mass

P Proportional band

U Heat-transfer coefficient

δ Decay ratio

σ_c Standard deviation of the controlled variable

τ_d Deadtime

τ_o Period of oscillation

τ_q Period of the load disturbance

τ_1 Primary time constant

Ω Overshoot

References

1. Kraus, T. W., and T. J. Myron, "Self-Tuning Controller Uses Pattern Recognition Approach," *Control Eng.,* June 1984, pp. 106–111.
2. Shinskey, F. G., *Process Control Systems,* 3d ed., McGraw-Hill, New York, 1988, pp. 354–356.
3. Kraus, T. W., "Pattern-Recognizing Self-Tuning Controller," U.S. Patent No. 4,602,326, July 22, 1986.
4. Luyben, W. L. (ed.), *Practical Distillation Control,* Van Nostrand Reinhold, New York, 1993, pp. 100–103.
5. Hang, C. C., and K. J. Åström, *Practical Aspects of PID Auto-Tuners Based on Relay Feedback,* Lund Institute of Technology, Lund, Sweden, September 1987.
6. Shinskey, *op.cit.,* pp. 52, 53.
7. Shinskey, F. G., "Adaptive pH Controller Monitors Nonlinear Process," *Control Eng.,* February 1974, pp. 57–59.
8. Albert, W., and H. Kurz, "Adaptive Control of a Wastewater Neutralization Process," IFAC Adaptive Control of Chemical Processes, Frankfurt, 1985.

Part

4

Nonlinear Elements

Chapter

8

Nonlinear Controllers

If the process being controlled is nonlinear, it is usually desirable to apply some type of complementary compensation to effect a linear closed loop. In this way, loop gain will tend to be constant over the expected ranges of load and controlled variable, allowing performance to be maximized across those ranges. Otherwise, controller tuning must be made to favor the regions of highest gain for stability reasons, resulting in sluggish control elsewhere. A classic example of this case is the pH control loop, where the controlled variable is a complex logarithmic function of the manipulated variable and load. A controller tuned to eliminate cycling responds very poorly to large load changes without nonlinear compensation, as shown in Fig. 7.7. However, this is only one reason for using a nonlinear controller.

In many cases, the process being controlled has such a low dynamic gain or the specifications on its controlled variable are so loose that on/off control is adequate. This allows the use of valve actuators that have only two states or manipulated variables that simply cannot be modulated. The use of single-speed bidirectional valve actuators also may be required for some applications. Still another possibility is a controlled variable that has only a limited number of states. Nonlinear controllers are needed for all these applications.

One additional reason for using a nonlinear controller is the possibility of achieving better control of a linear process than a linear controller can provide. This has prompted some manufacturers to add *fuzzy logic* to their PID controllers. All these cases are examined in this chapter.

Multiple-State Controllers

Under this heading are included all controllers used to drive multistate rather than continuous manipulated variables. For two-state devices, the on/off controller is used, although it has available enhancements such as pulse-duration modulation and derivative feedback as well. A three-state controller is required to drive a bidirectional single-speed motor, and the same enhancements can be applied to it. The important question with any of these multistate controllers is their performance when applied to a given process, which determines whether they can even be used or what enhancements may be necessary.

Two-state controllers

Two-state or on/off control is used commonly to manipulate variables that have only two states, such as solenoid valves, electric heating elements, home furnaces, humidifiers, air conditioners, etc. Other applications include filling a tank to a high level or pressure and allowing it to empty to a lower level or pressure by manipulating a single-speed pump or compressor. However, an on/off controller also may be used to manipulate a modulating control valve for processes having a very low dynamic gain, such as liquid level or pH in a flat portion of the titration curve.

Recognize, however, that the controlled variable in all these applications must limit-cycle if the load falls between the two states of the manipulated variable. For example, in heating a home, if the heat load represents a fraction of the supply capacity of the furnace, the temperature will be driven above set point when the furnace is on and fall below set point when it is off. This limit-cycling is usually acceptable at home but may not be in an environment where microprocessor chips are being fabricated. To make this determination, the amplitude of the cycle must be estimated relative to the dynamic properties of the process to be controlled.

Consider the case of an integrating process having a time constant τ_1 and a deadtime τ_d. Under on/off control, the controlled variable will ramp either upward or downward, with the rate determined by the difference between the present value of the manipulated variable and the load. If the load happens to be precisely centered between the two states of the manipulated variable, as it is in the first few cycles of Fig. 8.1, the cycle will be symmetrical and uniformly positioned about the set point. Any other load will distort the cycle and offset it, because up and down ramps then have different rates, as shown following the reduction in load from 50 to 30 percent in Fig. 8.1. The amplitude of the cycle $c_h - c_l$ remains the same for all loads between the output limits m_h and m_l:

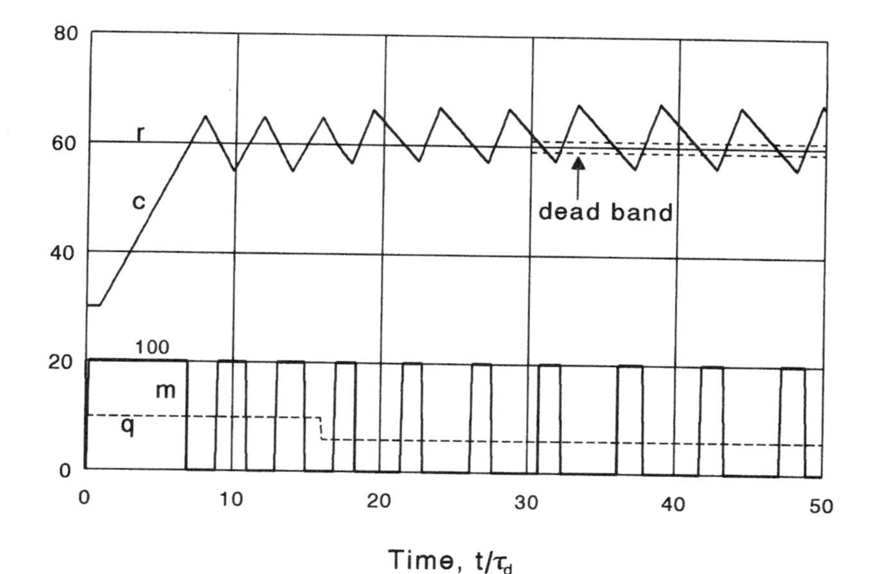

Figure 8.1 An integrating process under on/off control cycles symmetrically only at 50 percent load; adding dead band to the controller increases the amplitude by that amount.

$$c_h - c_l = (m_h - m_l)\tau_d/\tau_1 \tag{8.1}$$

The deadtime of the process in Fig. 8.1 is 10, and its time constant is 100, so the peak-to-peak amplitude of the cycle is 10 percent. The period of the *symmetrical* cycle is shortest, however, a point made in interpreting the results of closed-loop testing in Chap. 6. The period is a function of load relative to the output limits:

$$\tau_o = 4\tau_d\left(0.5 + 0.25\frac{q - m_l}{m_h - q} + 0.25\frac{m_h - q}{q - m_l}\right) \tag{8.2}$$

The preceding relationships apply strictly to the ideal on/off controller, i.e., one having no dead band. However, the ideal of having both on and off switching points identical is neither possible nor desirable. Any noise at all on the controlled variable would cause the output of the controller to chatter as the switching point were crossed. Introducing some dead band eliminates this sensitivity; the output switches on when the set point is exceeded in one direction by the dead band a, and off when exceeded in the other direction by the same value. The last four cycles in Fig. 8.1 were augmented by introducing a dead band of ± 1 percent into the controller, as indicated by the dashed lines spaced on both sides of the set point by that amount.

As observed, dead band increases the peak-to-peak amplitude estimated in Eq. 8.1 by $2a$. The dead band effectively augments the deadtime by $\tau_1(2a/100)$, where division by 100 is needed because dead band is expressed in percentage of full scale. With the corrections for dead band, Eqs. 8.1 and 8.2 are revised as follows:

$$c_h - c_l = 2a + (m_h - m_l)\tau_d/\tau_1 \tag{8.3}$$

$$\tau_o = 4(\tau_d + \tau_1 a/50)\left(0.5 + 0.25\frac{q - m_l}{m_h - q} + 0.25\frac{m_h - q}{q - m_l}\right) \tag{8.4}$$

If the process being controlled is *self-regulating,* then τ_1 should be divided by steady-state gain K_p where it appears in the preceding equations. Although this is an approximation, it is quite accurate in this context, because on/off controllers will be used primarily on lag-dominant processes where the amplitude of the cycle will be acceptably small.

In some applications, the dead band is used to extend the interval between starting and stopping motors that drive pumps, compressors, and machinery to minimize wear and heating. This is the case when filling reservoirs or pressure tanks with well water. The dead-band setting in this application should be as wide as allowable while taking into account the rising power and stress resulting from working against excessive pressure.

Three-state control

If two two-state variables require manipulation, a three-state controller is required, where in a center dead zone both manipulated variables are off. An example would be the application of either heating or cooling to a vessel, depending on the requirements of the current load and set point. Heating would be applied if the temperature fell below set point by z percent and cooling if it rose above set point by z percent: the dead zone would be $\pm z$. Without the dead zone, one or the other of the manipulated variables would be energized all the time, resulting in an excessive consumption of energy. To prevent this from happening, the dead zone must be wide enough that only one of the outputs is energized during a cycle.

Although Fig. 8.1 describes the case of only one manipulated variable, which drives the controlled variable upward, it can be used to estimate the behavior of the loop with a pair of manipulated variables. In responding to the initial set point, the controlled variable overshoots by 5 percent. The cycle is then uniformly distributed about set point at ±5 percent. Observe that decreasing the load from 50 to 30 percent offsets the cycle upward to +7 and −3 percent. Reducing the load to zero would therefore result in a 10 percent overshoot. If a second manipulated variable were to be added to drive the controlled variable downward, it should not be turned on until the deviation is 10 percent above the switching point for the first manipulated variable. As a minimum, then, the dead zone should be the peak-to-peak amplitude estimated using Eq. 8.1:

$$z + a \geq 100K_p\tau_d/\tau_1 \tag{8.5}$$

The value of 100 is used in Eq. 8.5 for the difference between the output limits under the assumption that the actual limits in a three-state application are 100 percent and 0. Note that dead band a also assumes part of the role of stabilizing the loop, as well as being useful to avoid chattering in the presence of noise.

It is entirely possible in this type of application for the load to be zero at times—at equilibrium, for example, where neither heating nor cooling is required. Then there will be no cycling, but the controlled variable is also free to float through the dead zone. In fact, the actual position of the controlled variable within the dead zone is load-dependent and therefore compares with the offset produced by a proportional controller. The two-state controller also produced offset in that the average value of the controlled variable was only at set point if the load was centered between the output limits. The principal difference in performance between two- and three-state controllers is that the latter does have a steady state in the load range covered by the manipulated variables, while the former does not.

This property is more prominent when a three-state controller is used to drive a bidirectional constant-speed valve motor. The three states are drive-open, stop, and drive-closed. Again, a dead zone must be provided that is wide enough to prevent the motor from cycling between the two directions and allow it to settle in a steady state. If the controlled variable is the position of a valve driven by the motor, there will be no deadtime, and the dead zone can be reduced very close to zero; some dead band also will be required. The process in the closed loop is an integrator represented by the valve motor, and τ_1 is the time required for it to travel full stroke.

Such a valve also can be used in a flow loop with a three-state flow controller. The response of this loop is simulated in Fig. 8.2, given a 2-second deadtime, steady-state gain of 1.5, and a stroking time of 100 seconds. (A similar result would obtain if the deadtime were 1 second and the stroking time 50 seconds, etc.) The dead zone is set at ±1.0 percent, as indicated by the dashed lines on both sides of set point, with a dead band of 0.5 percent at both edges. In responding to

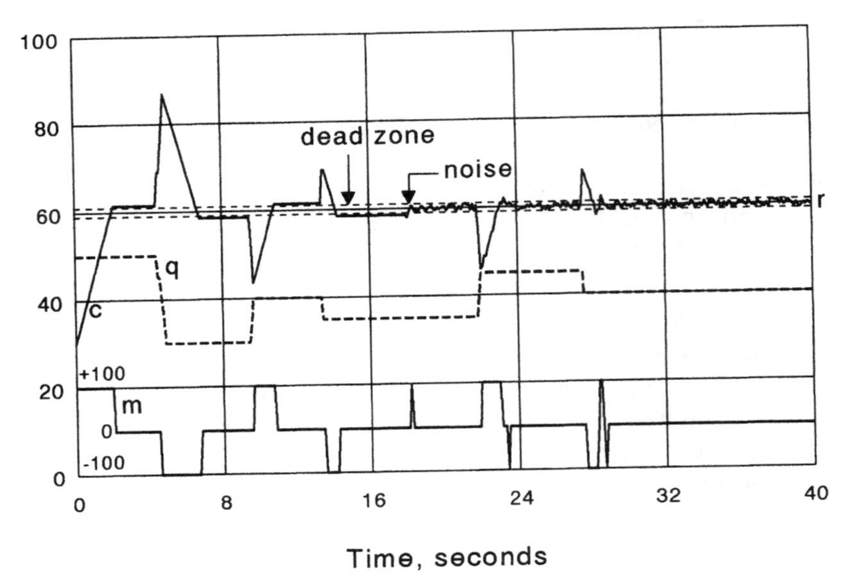

Figure 8.2 Controlling flow with a bidirectional constant-speed valve motor results in steady-state offset unless noise is present.

the initial large deviation, the flow overshoots set point by 1.5 percent and comes to rest. A large load change drives flow upward, and it is returned by the controller to a position 1.5 percent below set point. Subsequent smaller load changes produce identical results, the final state having a 1.5 percent offset. A load change of 2 percent or less could move the flow through the dead zone and dead band without initiating controller action, so any offset within ±1.5 percent is possible.

Then noise of 1.0 percent amplitude is added to the simulation, characteristic of flow measurements. This causes the dead band to be exceeded just enough for the motor to drive the flow to the center of the dead zone. Subsequent load changes also leave the flow centered, eliminating any measurable offset without any further valve movement. Noise is actually beneficial in this application. Its optimal level fills the dead zone completely; less would leave some offset, and more would require a larger dead band to avoid unnecessary valve motion.

Pulse-duration control

A two-state device can produce a proportional response if modulated on a duty-cycle basis. This is done by energizing it for a portion of a fixed time cycle. The period of the cycle must be well below the period of the loop, as determined by Eq. 8.2, or there will be no advantage

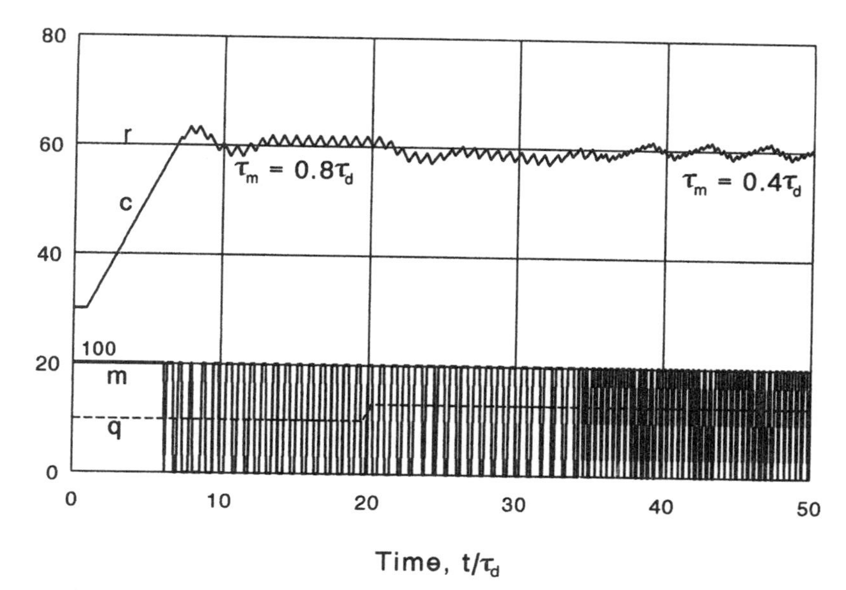

Figure 8.3 A pulse-duration controller applied to the same integrating process as in Fig. 8.1 produces a smaller amplitude cycle but is not free of offset or slower cycles.

gained. The *rangeability* of the device is the ratio of the minimum duration for which it can be energized to the period of modulation. And its *resolution* is the ratio of the scan period of the pulse-duration controller to the modulation period.

In Fig. 8.3, a proportional pulse-duration controller is applied to the integrating process of Fig. 8.1. A 10 percent proportional band is used, since the ratio of τ_d/τ_1 is 0.1. While the controlled variable is outside the proportional band, the manipulated variable is fully energized. However, as it passes through the proportional band, the percentage of time energized decreases, causing the controlled variable to settle near set point. The modulating period is initially set at 8, compared with the natural period of the loop at 40. As a consequence, the amplitude of the resulting cycle is 2 percent peak-to-peak compared with 10 percent where the on/off controller was used in Fig. 8.1.

Although the initial load and the controller's output bias are both 50 percent, the controlled variable settles out above set point. The modulated output matches the load perfectly, or the controlled variable would integrate upward or downward. However, the resolution of the manipulated variable is limited to 1/8, the ratio of the scan period to the modulation period. Hence the offset in this proportional loop could be ±6.25 percent, since anywhere in this range the duty cycle or percentage of time energized for the manipulated variable would not change.

A subsequent load increase of 15 percent shifts the controlled variable below set point but also produces a slower cycle, another product of limited resolution. The load is now 65 percent, but the nearest available duty cycles are 62.5 and 75 percent. This mismatch causes the controlled variable to integrate upward four times as fast as downward, but it will not rest. Later, the modulation period is reduced to 4, which reduces the amplitude of the short cycle in half but increases that of the long cycle. The latter effect is the result of the resolution being reduced to 1/4 so that the nearest available duty cycles are 50 and 75 percent. The natural period of 40 is now plainly visible.

The pulse-duration controller is a mixed blessing. In order to minimize the amplitude of the cycle in the controlled variable, the modulating period must be fast, but this reduces its resolution, producing a slow cycle. The most effective pulse-duration controller is one that has a very fast scan period relative to the natural period of the process. The ratio of 1/40 used in Fig. 8.3 was selected to illustrate its limitations. A modulating period of one-tenth the natural period is effective, but then the scan period needs to be one-hundredth the modulating period to give 1 percent resolution. Wear on the two-state device needs to be considered as a function of the modulating period. An electric heater driven by a solid-state switch can withstand fast cycling, but a solenoid valve with moving parts may wear out quickly, and electric motors will not tolerate frequent starts.

The pulse-duration controller is probably best applied as a modulator for a two-state device over a 0 to 100 percent range, accepting as an input a signal from a PID controller. In this way, the scan period for the pulse-duration controller can be much faster than that of the PID controller, as required for high resolution of the manipulated variable.

Adding proportional and derivative or integral action

Some room thermostats contain a tiny heating element connected to their output switch. When the switch activates the supply of heat to the room, it also sends power to this heating element, causing the temperature of the sensor to rise *before* that of the room. The thermostat then turns off the heat prematurely to avoid overshooting set point. In essence, there are two feedback loops present here: one through the process and one through a "model" of the process—the small heating element in the thermostat case. The effect is similar to that of a Smith predictor, where a fast negative-feedback loop lacking any deadtime, in parallel with the process, allows the use of a high-gain controller. However, this practice predates Smith's invention and possibly even the addition of derivative action to a pneumatic controller, which it resembles even more closely.

The early pneumatic controllers achieved proportional action by feeding back the output signal through a bellows to the deviation linkage so that the output was proportional to the deviation; the proportional band was set by the ratio of the lengths of the feedback and deviation levers. Later, a restrictor was placed at the inlet to the proportional feedback bellows. This lagged the feedback signal, producing a higher gain at high frequencies compared with the proportional gain at steady state. Derivative action was the result, whose time constant was that of the lag produced by the restrictor and its capacity.

The same concept can be applied to a two- or three-state controller, as shown in Fig. 8.4. The two- or three-state output is fed back to the controller input through a first-order lag and adjustable gain. Consider first its application to a two-state controller having a single output. It would be governed by the following digital code:

```
IF e + (f - b) * P / 100 > a THEN m = ml
IF e + (f - b) * P / 100 < -a THEN m = mh
```

where f is the output of a first-order lag having a time constant of D and a scan period of dt:

```
f = f + (m - f) / (1 + D / dt)
```

This describes a PD controller having adjustable proportional, derivative, and dead-band (a) settings and output bias (b). The output (m) cycles between high (mh) and low (ml) limits.

Figure 8.5 describes the results of applying this controller to the same integrating process controlled in Figs. 8.1 and 8.3. A 10 percent proportional band was used, and a derivative time of 0.3 deadtimes was found to be optimal; dead band was set at 0.5 percent. The output

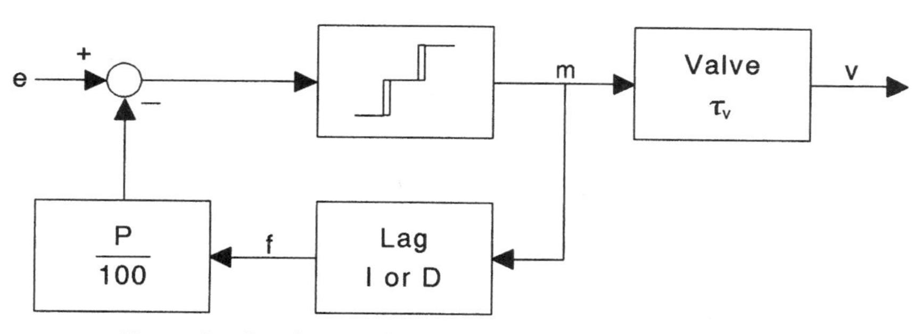

Figure 8.4 Proportional and integral or derivative action can be added using a feedback loop from the contactor output.

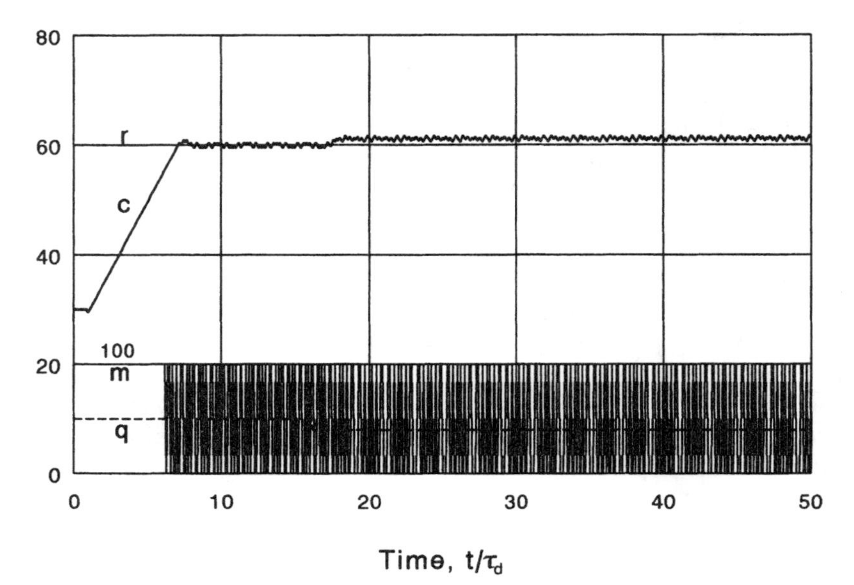

Figure 8.5 Proportional and derivative actions are here added to the two-state loop of Fig. 8.1, producing an effect similar to pulse-duration control, but tighter.

motion of the controller is similar to that of the pulse-duration controller, but the performance is significantly improved owing to the derivative contribution, with very little set-point overshoot. A slightly negative offset appears when the load equals the bias at 50 percent and about a 1.5 percent positive offset after the load is reduced to 40 percent.

If applied to a three-state controller having two two-state outputs, a dead zone would have to be added and the bias set to zero, which would represent the middle of this load range, just as 50 percent represented the middle of the load range for one output. If the outputs are connected to opposite windings of a bidirectional valve motor, then that motor participates as an integrator in the control. The lag in the feedback path of the controller then becomes the integral time I, and the true proportional band is $P\tau_v/I$, where τ_v is the stroking time of the valve. The motorized valve has limit switches at each end of its travel to turn off the driving winding when its limit is reached. Tests were made with the feedback signal to the lag taken both upstream and downstream of the limit switches to observe behavior coming out of saturation. The upstream connection as in Fig. 8.4 is preferred, in that PD action is obtained with the valve at a limit, avoiding overshoot when the loop recovers from the limited condition.

Dual-Mode Control

"Bang-bang" control has been favored for some time in the control of vehicles, in that it can achieve the desired positioning of a vehicle in either minimum time or using minimum energy with very predictable results. In essence, full power is applied until a certain position is reached, at which point power is turned off or full braking is applied just long enough to bring the vehicle to the designated position. However, vehicle positioning is considerably simpler than process control, in that the load is usually zero and there is little, if any, deadtime in the loop. Lags are usually inertial and can be estimated by comparing force applied against acceleration and velocity produced.

This technology has been applied successfully to process control, but with more complications and limitations. The complications arise from there being a variable and unpredictable load. The controlled variable cannot simply be brought to the new set point and the power turned off or the brake locked. The manipulated variable must be adjusted continuously to match the variable load in order to maintain equilibrium at set point. As seen in Figs. 8.1, 8.3, and 8.5, this cannot be accomplished with a two-state device, because the load does not match either state. Hence "bang-bang" control cannot be used to maintain a steady state, but only to drive the process from one steady state to another. A modulating controller such as a PID controller must be coordinated with it to maintain a steady state. The combination of the two methods is known as *dual-mode control.*

Time-optimal control

The "bang-bang" part of dual-mode control is called *time optimal* in that its function is to produce the desired result in minimum time; a subclass produces the result using minimum energy. It is three-state control where the center state is intended to hold the controlled variable constant. In process control, this requires the manipulated variable to match the load. To accomplish the stated objective, switching must take place from acceleration to deceleration at a precise deviation short of set point and deceleration held for a precise time delay. For a non-self-regulating process following an increasing set point, the deviation e_l and time delay t_l are

$$e_l = [(m_h - q)(\tau_2 + \tau_d) - (q - m_l)t_l]/\tau_1$$

$$t_l = -\tau_2 \ln [(q - m_l)/(m_h - m_l)] \qquad (8.6)$$

where m_h and m_l represent the highest and lowest states of the manipulated variable relative to the load q. A similar pair of equa-

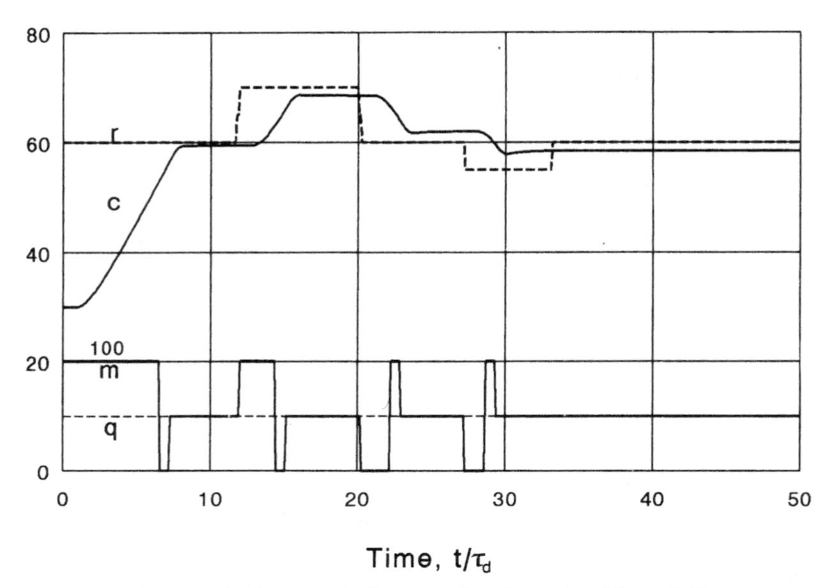

Figure 8.6 Time-optimal control of a second-order non-self-regulating process with deadtime results in some offset, which increases for smaller set-point changes.

tions is required to estimate the switching point and time delay for a decreasing set point:

$$e_h = [(q - m_l)(\tau_2 + \tau_d) - (m_h - q)t_h]/\tau_1$$

$$t_h = -\tau_2 \ln [(m_h - q)/(m_h - m_l)] \tag{8.7}$$

Figure 8.6 shows a simulation of this time-optimal system applied to a non-self-regulating process having a deadtime τ_d and secondary lag τ_2 each equal to one-tenth the primary time constant τ_1 at a load of 50 percent. For the initial set-point change, there is a small undershoot, probably caused by limited resolution in time; switching can only take place at regular sample intervals of the controller. For this simulation, there are 10 sample intervals to the process deadtime, and the time delay happens to be 7 sample intervals long. Faster sampling would be required for more accurate results—analog systems do not have this limitation.

The offset is greater for the smaller set-point changes because the velocity of the controlled variable at the point of switching has not reached as high a level as it did during the large set-point change. This could be corrected by adding derivative action to the algorithm, which would adjust the switching point relative to the velocity of the controlled variable. However, since derivative action amplifies noise, switching could then become erratic. The last set-point change produced no control action whatever because the deviation did not exceed the switching point, which in this example is ±6.5 percent. This experience indicates that time-optimal control is primarily useful for responding to large set-point changes, to the exclusion of small set-point changes, and load changes as well, which begin as small deviations.

Minimum-energy control avoids switching from acceleration to deceleration, allowing the controlled variable to coast to a stop. The time delay in this case is set to zero, resulting in a larger deviation at the switching points. If the process contains no secondary lag, there is also no time delay, and the same result obtains.

Application to batch reactors

A very demanding process requiring both exceptional set-point response and load rejection is the batch exothermic reactor. It tends to be steady-state unstable, having negative self-regulation, as described in Chap. 2. If the controller is not tightly tuned, a thermal runaway could develop, destroying a batch of product and possibly endangering equipment and personnel. Integral action is essential, since offset cannot be tolerated, nor can cycling. Large set-point changes are common, and overshoot must be minimized, yet the time to reach set point also must be minimized because it represents lost production capacity. A cascade configuration as shown in Fig. 7.9 is preferred to speed response of the primary loop and reject secondary-loop disturbances. Tuning this system for load response is no problem. Achieving fast set-point response without overshoot is. Various techniques have been tried with some degree of success, including dual-mode control.[1]

The reactor temperature is raised by applying full heat. To minimize time and avoid overshoot, full cooling must be applied before set point is reached, based on the deadtime and secondary lags in the loop. The switching point can be set at a designated deviation below set point that is a function of the deadtime and secondary lags and the rate of temperature rise. Even if the deadtime and secondary lags are well defined, the rate of temperature rise is quite variable:

$$\frac{dT_r}{dt} = \frac{Q_r - UA(T_r - T_j)}{MC} \tag{8.8}$$

where T_r is the reactor temperature, T_j is the jacket temperature, Q_r is the heat produced by the reaction, MC is the product of reaction mass and heat capacity, and UA is the product of heat-transfer coefficient and area. During the rise toward set point, both temperatures are variable, and so is Q_r, which varies exponentially with T_r. Furthermore, MC and UA may vary from batch to batch.

When full cooling is begun, the temperature difference begins to fall and eventually reverses, at which point full cooling may be discontinued in that the reactor temperature should stop rising. Once the set point is reached, however, the steady state must be maintained by holding dT_r/dt to zero, which requires balancing the temperature difference against the unknown value of Q_r:

$$T_j = T_r - Q_r/UA \tag{8.9}$$

The PID controller must be "preloaded" with this estimated value of T_j to avoid bumping the loop when control is transferred to it. (In the simulation of Fig. 8.6, the middle state of the controller was set equal to the load, which was known and constant.)

In the first application of this technique to batch reactor control, described in ref. 1, the system was tuned for nearly perfect set-point response at the reaction temperature for the polystyrene recipe being used. However, after the exothermic reaction was complete, the set point was then elevated to cure the product. Not being tuned for this higher set point and lack of heat production, neither the switching point nor preload were correct. This resulted in premature cooling and transfer to PID control, with time wasted approaching set point, followed by overshoot. Later modifications included multiple settings for various operating conditions and even an attempt to adapt settings as a function of set point and coolant temperature. As the variety of recipes used in a particular reactor increased, new combinations of settings had to be added for each. There remained the unknown condition of the heat-transfer surface that could change gradually with time; when the settings given with the recipe no longer produced acceptable control, the vessel needed cleaning.

I spent considerable time attempting to develop an adaptive dual-mode system for this application, with less than satisfactory results. The unknown values of Q_r, MC, and UA all needed to be determined based on the rate of temperature rise while heating the reactor, and Q_r could not be estimated accurately in the unsteady state. It changed too much with temperature, doubling every 10°C, and the lags in the temperature measurement left the estimate too far behind. The effort was finally abandoned in favor of the PID cascade system with secondary temperature feedback to the primary integral mode described in connection with Fig. 7.9, which does not require retuning as operating conditions change.

Multiple-State Inputs

There is a small class of control loops having the opposite characteristics of those described above. Instead of having a high-resolution mea-

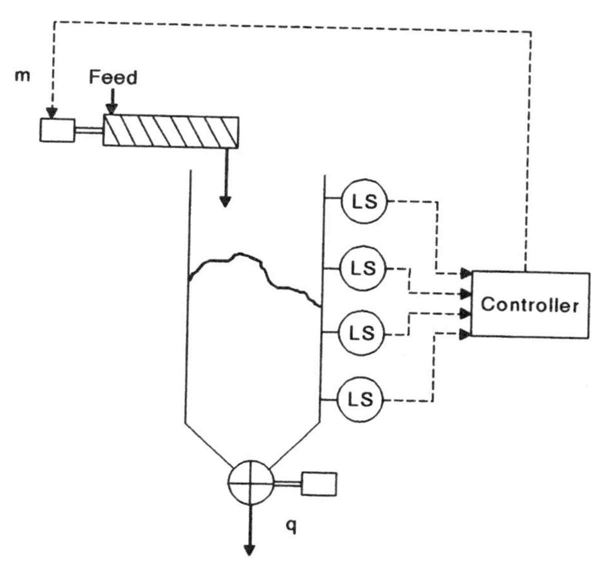

Figure 8.7 The level of solids in the hopper is sensed only at four points, yet a smooth inflow must be delivered.

surement of the controlled variable and a manipulated variable with two or three states, we are presented with a multiple-state controlled variable and a high-resolution manipulated variable. A simple example appears in Fig. 8.7. A mill is drawing a solid material from a hopper at a variable rate, and its inventory must be controlled by manipulating inflow. The problem is that the inventory cannot be measured continuously but only detected at four points in the hopper. Preferably, the level should be kept between the second and third detectors, with the first and fourth only required for alarms, shutdown, and startup.

This is not an easy problem to solve. If proportional control were to be used, the manipulated inflow would have five states—0, 25, 50, 75, and 100 percent—related to the five known states of the level of solids in the hopper. The load is not likely to match any one of these inflow states, causing the inflow to cycle between the two states on either side of the load and the level to cycle between the two corresponding positions. This may be satisfactory, or it may not.

An integrator can be used in an effort to take advantage of the full resolution of the inflow. It would change the rate of inflow as a function of the time that the level has been in a state. In the center state, inflow would be held constant, while on either side of center, inflow would be increasing or decreasing until the level returned to the center state. In the two outside states, the rate of change would be faster. Unfortunately, this loop also will cycle, because the process is non-self-regulating; two integrators in a closed loop form an oscillator. To deliver a smooth inflow without cycling requires a more complex system.

Using models

If the load is measurable, e.g., by the speed of the star feeder at the bottom of the hopper, then the speed of the feeder at the inlet can be set proportionally; this is feedforward control. However, since the process is non-self-regulating, any mismatch between the two flows will cause the level to ramp either up or down. Proportional control can be added to change inflow stepwise by a given increment when the level leaves the center zone, removing the increment when it returns. The loop may still cycle, but the period can be extended considerably if the feedforward calculation is accurate and the proportional increments can be made correspondingly small.

Lacking the feedforward input, the load can be estimated based on the time required for the level to return to the center state following the application of an incremental change in inflow. Consider the case where the level leaves the center state and a proportional change Δm_1 is applied to the inflow to bring it back. Because of some dead band in the level detector, the level must change by an amount Δh before the detector changes state. The time required for this level change is

$$\Delta t = \frac{\Delta h \, \tau_1}{m - q} \tag{8.10}$$

where τ_1 is the integrating time of the vessel, and m and q are present values of the manipulated variable and the unknown load. This difference between m and q must be set to zero by incrementing m an amount equal and opposite to the difference:

$$\Delta m_2 = -(m - q) = -\Delta h \, \tau_1 / \Delta t \tag{8.11}$$

As with the feedforward system, the accuracy will not be perfect, and some slow cycling will result. The larger the dead band in the level detectors, however, the more accurate the estimate of load will be because both Δh and Δt will increase; the sensitivity to noise also will be reduced thereby.

Using fuzzy logic

At this writing, fuzzy logic is being hailed as a momentous advance in process control, allowing automation of processes that resisted other methods. For example, ref. 2 reports that "...demethanizer and propylene columns have long been impossible to automate using conventional controllers," and "fuzzy control has allowed us to fully automate them." The first of these statements is simply not true, feedforward material-balance control having been applied successfully to both these columns in many installations around the world. These claims had to have been made by people who were unaware of more conventional methods for solving the same problems. This conclusion is confirmed by another quotation from the same news item:

> ...using a conventional PID controller to achieve the same results [as fuzzy logic] would require very specific rules: "If the water temperature is between 130°F and 150°F and the water level is between 12 cm and 20 cm, then the flowrate of cold water must be decreased by 20%."

This is *not* a description of a PID algorithm, but rather of a multistate algorithm such as some described in this chapter, and it would result in cycling similar to that shown in many of the figures appearing here.

This is not an isolated example. In ref. 3, a similar misrepresentation of PID control appears:

> ...[in contrast to fuzzy logic] a rule in a conventional proportional-integral-derivative (PID) controller would need to be very specific: If brake temperature is greater than 280 and speed is less than 45, then brake pressure is 190.

The inequalities and single output state identify this as a multiple-state algorithm.

In a letter to the editor commenting on ref. 3, a colleague, Edgar Bristol, pointed out that "fuzzy logic does not repeal the laws of feedback and stability." In other words, there are limits to both the performance and robustness of feedback loops, as outlined in Chap. 2, that cannot be violated. If fuzzy logic is to improve on the performance of a feedback loop, it must do so within those limits. There is no magic here; fuzzy logic is a combination of algebra and multiple-state logic, which is why it is covered in this chapter.

Probably the earliest application of fuzzy logic was in the control of cement kilns. This is a multivariable process where some of the factors affecting product quality cannot be measured on-line, and quality must be reported after the fact from laboratory analysis. The variable having the greatest effect on quality is the temperature of the "burning zone" in the kiln. Operators may only be able to estimate it visually as "too high," "just right," or "too low." In observing other variables, the operators were found to make similar judgments and combine their judgments into actions such as "increase fuel flow 2.5 percent." Not all operators were equally proficient in producing quality product with minimum fuel, so an attempt was made to convert these judgment calls into a set of rules that could be executed by a computer and adjusted to optimize performance. To implement the operators' rules, it became necessary to make a computer process information like a human being—hence the title of ref. 4, "Making Computers Think Like People."

Multiple-state logic has the disadvantages noted above, low resolution producing cycling, etc. Fuzzy logic smooths the transitions between these states by giving each a sigmoidal membership distribution[4] and adding the contributions of each where they overlap. In practice, the sigmoidal curves are usually approximated as straight lines for convenience.[5] In a multivariable process such as the cement kiln, the various inputs, after determining their membership, are weighed on the basis of their relative effect on the quality of the product and are combined into a calculated adjustment to the manipulated variable(s). In essence, a model is used to make this estimate—it is not the usual mathematical model based on first principles, but it is a model nonetheless. The use of fuzzy logic produced an improvement in the operation of cement kilns, but before it, there was no feedforward model nor feedback from a product-quality measurement, so the improvement should not be surprising.

Now, various controller manufacturers are incorporating fuzzy logic into their single-loop PID controllers,[5] claiming improved performance. In a single loop having an accurate measurement of the controlled variable, e.g., temperature, the reasons found in the cement kiln application do not apply. Instead, it is being used to depress set-point overshoot and in self-tuning. For example, the Yokogawa UT35 controller shifts the set point toward the controlled variable when the velocity of approach is likely to result in overshoot. Within a window around zero deviation, the fuzzy logic has no function. However, if the deviation exceeds a specified level and/or its derivative exceeds a similar threshold, rules are invoked that depend on the sign and direction of the deviation. The Omron E5AF controller similarly uses fuzzy logic to suppress set-point overshoot.

As described in Chap. 6, there are other methods for minimizing set-point overshoot using different tuning rules, a set-point filter, or a model-based controller. In fact, wherever fuzzy logic has been used, there are other methods available to achieve the stated objective with a higher level of accuracy or better performance. Fuzzy logic is an approximation to mathematical modeling and therefore cannot be superior to it. The fact that it can interpolate between states gives no advantage over precise calculation of the true value of the variable in question. In developing rules for self-tuning a PIDτ_d controller, I found it necessary to use very precise calculations of the tuning parameters because of the limited robustness of the control loop, and fuzzy logic did not have the required accuracy. At a time when 32-bit processors are state of the art, fuzzy logic processors have only 4-bit resolution.

Arguments that fuzzy logic brings powerful nonlinear features to the control loop are similarly overstated. The nonlinear functions described in this chapter can all be produced without it by using combinations of multiple-state logic and mathematical relationships. Furthermore, the performance of linear controllers on linear processes is superior for load changes and small set-point changes. Large set-point changes can benefit from time-optimal control, as described above, although the benefits are difficult to realize over a wide operating range.

Nonlinear PID Controllers

Under this heading is included PID controllers to which nonlinear characterization has been added in an attempt to alter their behavior to favor one operating objective or condition over another. Nonlinear characterization was discussed in relation to pH control in the preceding chapter, but in that context, its purpose was to linearize a nonlinear loop. The present focus will be on the intentional addition of a nonlinear function to an otherwise linear loop. For example, the averaging-level control application discussed in Chap. 3 needed to keep manipulated feed rate to a process as constant as possible without overflowing or emptying the feed tank. Various nonlinear level controllers have been tried in an attempt to solve this problem.

Error-squared controllers

One of the controllers tried for averaging-level control is the error-squared controller. Its proportional gain is made proportional to the absolute value of the deviation, causing its output to be proportional to the square of the deviation, with the sign of the deviation preserved. While the deviation is at zero, the controller has zero gain, which seems to be desirable for averaging level control, but since the process is non-self-regulating, any mismatch between the current load and a constant manipulated outflow will cause the level to move away from zero deviation until enough feedback is applied to achieve a balance. As a result, the deviation never comes to rest at zero. In fact, without any gain at zero deviation, the level will simply cross through on the way to the other side, culminating in a slow oscillation.

A significant danger in this and other applications is that the controller will be tuned only under conditions of small deviations, in which case the loop may not be stable for large upsets. Conceivably, a large disturbance could produce a deviation large enough to cause the loop gain to exceed 1.0. This would begin an expanding cycle that could continue to expand because of increasing controller gain until output limits are reached or some damage is done to the process. In

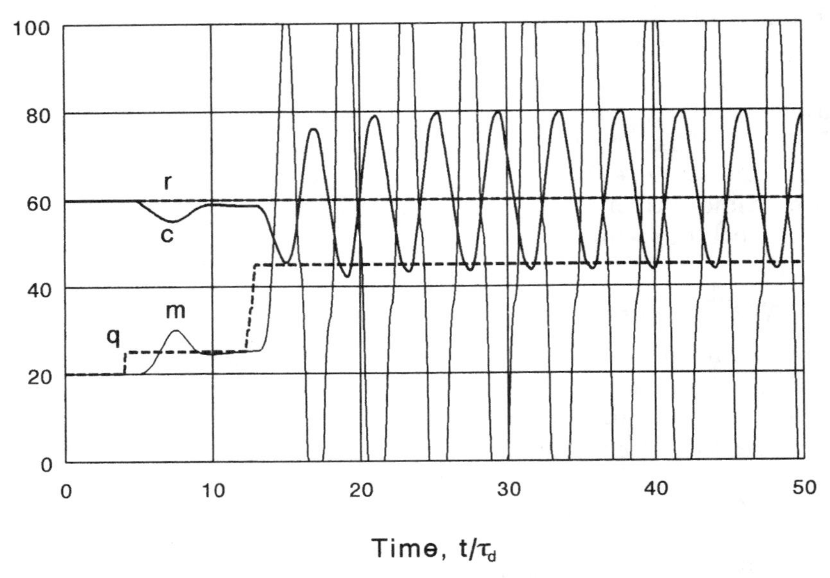

Figure 8.8 If the error-squared controller has too low a proportional band, the loop can go unstable for large upsets even when it is stable for small ones.

the wrong hands, this controller could be an accident waiting to happen. Figure 8.8 shows the results of setting the proportional band too narrow on the error-squared controller of a liquid-level loop. A 5 percent load change produces a well-damped response and even some apparent offset (it is this tendency to control with offset that encourages users to reduce the proportional band to dangerous levels). Next, a 15 percent load change precipitates an expanding cycle, from which the loop cannot recover without manual intervention.

Reference 6 describes an error-squared controller whose integral time I changes with deviation, as does its proportional gain K_c, to keep the product K_cI constant. This was intended to maintain a constant damping factor for averaging level control. I attempted to use one of these controllers on the liquid level in a kettle reboiler of a distillation column. The measurement was quite noisy, and the controller had the capability of filtering the noise with its error-squared function. However, shortly after installation, an upset arrived that was large enough to cause an expanding cycle; the controller was immediately replaced with a linear one before the operators lost confidence.

A shaping adjustment L can be added to the error-squared controller to increase its minimum gain and diminish its nonlinearity:

$$K_c = \frac{100}{P}[L + (1 - L)|e|/100] \tag{8.12}$$

where L is essentially the controller's linearity, adjustable over a range of 0 to 1. Set at zero, the controller is error-squared, and at 1, it is linear, with intermediate values giving the curve intermediate shapes. One of these controllers was originally developed by me in 1961 and applied successfully to the control of noisy flow. It was found to be not sufficiently adjustable for pH applications, however, which generally required a larger region of low gain around zero deviation. This ultimately led to the segmental nonlinear controller, applied to pH in 1963.

Segmental nonlinear control

The three-segment nonlinear function used in this controller is overlaid on a titration curve in Fig. 7.8 to illustrate its use in pH control. In the present context, however, the function is examined for possible application to linear processes. It is less likely to be mistuned than the error-squared controller above, in that its gain does not continuously increase with deviation but has essentially only two values. Still, it is not recommended for liquid-level control, in that the level tends to drift through the low-gain zone, controlling at one or the other corner where the zones meet. Most of these controllers have been removed from averaging-level applications as being ineffective; proportional control as described in Chap. 3 is recommended.

However, this nonlinear function is valuable as a noise filter; the width of the low-gain zone should be set to match the amplitude of the noise, and its gain should be set between 0.1 and 0.2 to be effective. The noise is attenuated by this factor as long as the process remains in control, i.e., as the deviation averages zero. However, even a small disturbance will cause some excursions into one of the high-gain zones, causing the controller to respond with corrective action. Figure 8.9 shows the simulation results for a liquid-level loop such as on a kettle reboiler, with 1 percent noise on the controlled variable. The controller has a proportional band of 50 percent, so the noise on the output is doubled. The nonlinear filter is not used for the first half of the test. Then the low-gain zone is set equal to the noise level, as indicated by the dashed lines, with a gain of 0.1. Noise on the output diminishes substantially until the following load change forces the deviation out of the low-gain zone. Once it returns, the noise almost disappears, yet the response of the loop to the load change is virtually unaffected by the filtering. It is as responsive as in the linear case but damped more heavily because of the low gain at set point. The key to keeping this controller's performance high is keeping the low-gain zone no wider than the noise level.

This type of filter was used to reduce the derivative spikes to an acceptable level in the temperature-on-temperature cascade configu-

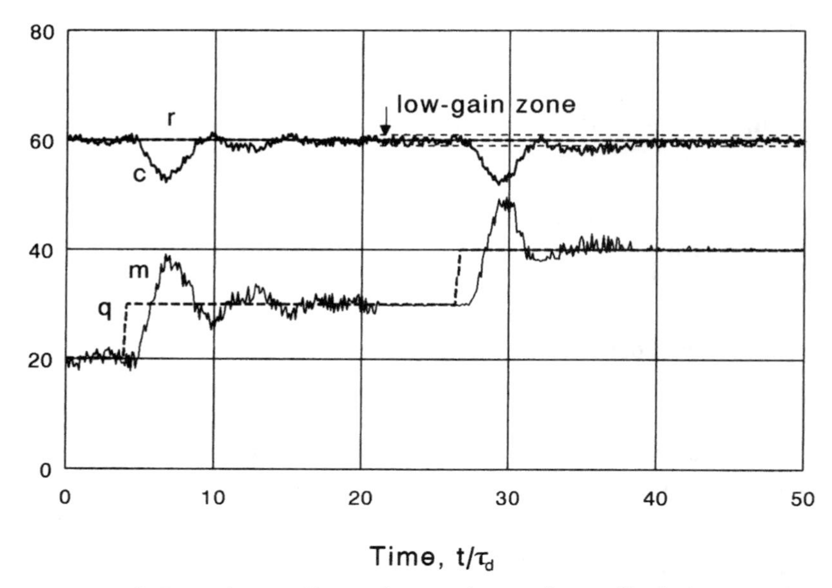

Figure 8.9 A three-piece nonlinear characterizer performs effectively as a noise filter without sacrificing load response.

ration shown in Fig. 7.9 and described at the end of Chap. 4. It had to be used in both controllers to provide the required amount of filtering with minimal loss of primary-loop performance. Adding a lag filter between the controllers reduced the spikes effectively but at the price of extending the period and dynamic gain of the primary loop to an unacceptable degree. The lag affects performance at all deviation levels, whereas the nonlinear filter does so only in the vicinity of zero deviation.

Output characterization

As mentioned briefly in the preceding chapter, the characteristics of the control valve manipulated by the controller often do not match those required by the process over the load range. When this happens, the controller must be tuned for conditions where the loop gain is highest to provide stability there, resulting in sluggish performance elsewhere. Proper compensation ought to be provided in every case to minimize this problem.

There are three different methods in use to compensate valve characteristics, one of which applies a nonlinear valve positioner. A valve positioner compares the position of the valve stem against the controller output signal and applies force to make them match. Some models contain a cam placed between the stem and the controller output, allowing the insertion of some nonlinear relationship between them. The cam may be contoured either to make an equal-percentage valve respond linearly or a linear valve assume an equal-percentage characteristic. There are turndown limitations to the cam, and producing a gain change of 50 to 1 may be too much to expect.

A second method uses a segmental characterizer inserted between the controller and the valve to produce the same result. These characterizers consist of a number of straight-line segments connected to approximate the required curve. Twelve segments are probably adequate for valve characterization. If a linear valve is to be shaped to a particular nonlinear function, the characterizer is simply matched to that function as smoothly as possible. However, if a nonlinear valve is to be linearized, the characterizer must take on a shape opposite to the valve characteristic. In this case, the valve characteristic must be plotted using linear coordinates on a transparency and the sheet turned over so that input and output axes are reversed. The characterizer must then be fit to the resulting curve.

A third method fits a hyperbolic function to the required characteristic. The function of choice is

$$f(m) = \frac{m}{L + (1 - L)m} \tag{8.13}$$

where $f(m)$ is the output of the characterizer, and m is the output of the controller; adjustable parameter L sets the linearity of the function, just as it did with the error-squared controller above. If L is set to 1, the function is linear, whereas greater or lesser values shape the curve in opposite directions, with all curves passing through (0, 0) and (100, 100) percent. Figure 8.10 shows a sampling of the curves it can generate for values of L between 0.1 and 10. Values greater than 1 can be used to make a linear valve appear to be equal-percentage, and values less than 1 can be used to make an equal-percentage valve appear to be linear. (Valves that close on an increasing signal require the opposite characterizations.)

Because these curves are hyperbolic rather than exponential, they cannot compensate perfectly for an equal-percentage (logarithmic) valve, yet a surprisingly close fit can be obtained. Figure 8.11 shows the characteristic of an equal-percentage valve having a rangeability of 50, both with and without hyperbolic compensation. A linearity setting of 0.2 in the compensator forces the compensated curve almost through the (50, 50) point. The curve is out of linear only about 3 percent, entirely adequate for feedback control. Valves having higher rangeability also have more gain change, causing the compensated curve to show more distortion. Yet compensation has been satisfactorily applied using this method even up to a rangeability of 1000.

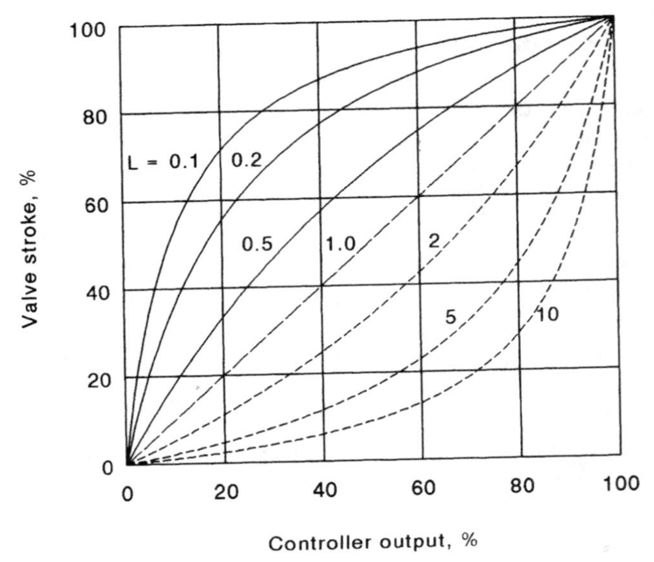

Figure 8.10 The hyperbolic function serves as a universal valve characterizer.

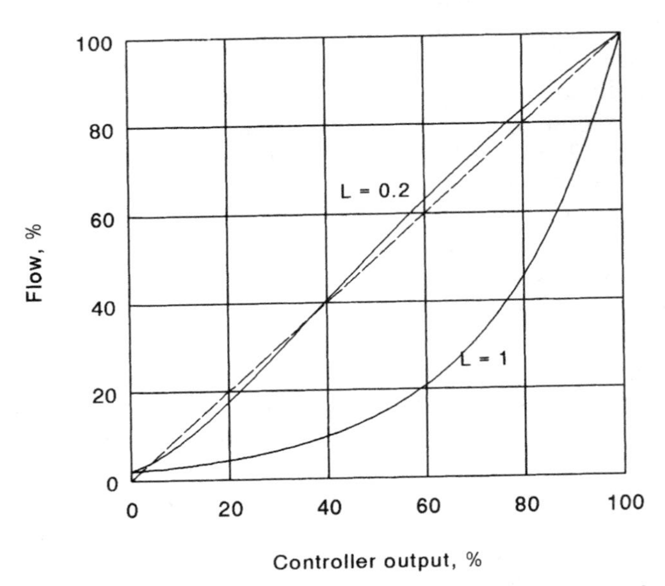

Figure 8.11 An equal percentage valve having a rangeability of 50 can be linearized quite satisfactorily using the hyperbolic function.

Notation

a, a	Dead band
A	Heat-transfer area
b	Output bias
c	Controlled variable
c_h	Higher value of c
c_l	Lower value of c
C	Heat capacity
d	Derivative operator
D, D	Derivative time
dt	Sample interval
e, e	Deviation
e_h, e_l	Deviation at which switching takes place
f, f	Function, feedback signal
h	Level
K_c	Controller proportional gain
K_p	Process steady-state gain
I	Integral time
L	Linearity adjustment
m, m	Manipulated variable
m_h, mh	High limit of m
m_l, ml	Low limit of m
M	Mass
P, P	Proportional band
q	Load
Q_r	Heat production rate
r	Set point
t	Time
t_h, t_l	Time delays following switching
T_j	Jacket outlet temperature
T_r	Reactor temperature
U	Heat-transfer coefficient
z	Dead zone
Δ	Difference
τ_d	Deadtime
τ_m	Modulation period
τ_o	Period of oscillation

τ_v Valve stroking time
τ_1 Primary time constant
τ_2 Secondary time constant

References

1. Shinskey, F. G., and Weinstein, J. L., "A Dual-Mode Control System for a Batch Exothermic Reactor," 20th Annual ISA Conference, Los Angeles, October 4–7, 1965.
2. Samdami, G., "Newsfront: Fuzzy Logic, More Than a Play on Words," *Chem.Eng.*, February 1993.
3. Cox, E., "Fuzzy Fundamentals," *IEEE Spectrum,* October 1992, pp. 58–61.
4. Zadeh, L. A., "Making Computers Think Like People," *IEEE Spectrum,* August 1984, pp. 26–32.
5. Infelise, N., "A Clear Vision of Fuzzy Logic," *Control Eng.,* July 1991, pp. 28–30.
6. Shunta, J. P., and Fehervari, W., "Nonlinear Control of Liquid Level," *InTech,* January 1976.

Chapter

9

Constrained Operation

At this writing, there are three competing programs attempting to arrive at a universal Fieldbus standard governing the features of "interchangeable" controllers at the lowest level in process plants. While begun as a combined effort of the Instrument Society of America (ISA) and the International Electrotechnical Commission (IEC), delays in reaching a consensus caused a group of manufacturers to form the Interoperable Systems Project (ISP). Not long afterward, another group of manufacturers formed WorldFIP. All three groups are dedicated to the single purpose of producing a standard for digital control products that would allow a field controller to be replaced by another from a different manufacturer without any operational difference which would affect either control performance or operator access.

While this is an admirable goal, standards have a way of killing innovation and stifling development. They also tend to eliminate product differentiation. If all controllers are operationally identical, users will tend to select the cheapest, and premium devices or premium features will have no value. To avoid disqualifying any member of the group, the standards will have to reflect a minimum combination of features rather than a maximum. As a result, there is competition even within a group over the standard features, since each manufacturer would prefer to see its own features as the standard. And although there is little argument over such specifications as pertain to the PID modes themselves, disputes have arisen over the issues raised in this chapter, i.e., over the features of constrained operation.

A principal issue facing all the groups is integral-windup protection. The chairman of the ISA User Application Layer Subcommittee, Richard Lasher, has his own method of windup protection which uses logic based on whether the final actuator has reached one of its limits; as a result, the indicators of a limit having been reached have taken on the name *Lasher bits*. There are competing methods for windup protection which are in some applications simpler to implement or more effective or both and which are presented in this chapter. This is but one example of the controversy surrounding the Fieldbus standards. There are others as well, related to the number of bits available to signal the status of controllers, inputs and outputs, and other function blocks, as well as how the information is passed. Needless to say, these considerations are important in determining how a control system responds to startup, off-limits operation, and shutdown conditions, as well as operator intervention.

Auto-Manual Transfer

Most controllers may be placed in manual operation, in which state the operator may adjust the output to any desired value. Transferring to manual must be recognized as the most expeditious way to stop an oscillation in a control loop, and most operators are not timid about making the transfer whenever there is a suspicion that a controller is doing more harm than good. After the trouble has passed, the operator should return the controller to automatic as quickly as possible. Therefore, it should be as easy to transfer one way as the other.

Bumps during transfer

Early pneumatic controllers such as the large-case instruments of the 1940s used the air-supply regulator for manual output and a simple two-position transfer switch. This minimized parts but could easily bump the process upon transfer. For example, if the controller were transferred to manual using the existing setting of the regulator, the output to the valve would step to maximum. However, the operator could override the automatic output simply by reducing the supply pressure to any desired level. The regulator acted as a high limit on the output, but if the controller were left in the automatic mode, it could reduce its output below that limit, so transfer to manual was still necessary if this action were to be avoided. Transfer back to automatic produced an output no higher than the existing manual output, but it could be lower, depending on the deviation at the time of transfer. The controller could not wind *up* while in manual, but it could wind *down,* again depending on the deviation existing while in manual. However, since final actuators are always (or should be always) selected to fail safe on a loss in signal, a lower output is always safer than a higher one, so the bump that was possible on a transfer back to automatic was at least not hazardous. The transfer to manual could be hazardous, however, if the transfer were made prior to reducing the output of the supply regulator.

Later pneumatic controllers had separate supply and manual-output regulators; the supply regulator was externally mounted, and one could be used to supply several controllers. This allowed the manual regulator to be left at what was considered a safe or standby setting, because when the controller was transferred to manual, its output immediately went to that setting. The resulting bump may have been desirable for some applications but not for most; in most cases, the operator preferred a bumpless transfer to manual. While the controller was in manual, the control loop was open, and any deviation appearing at its input would be integrated, causing windup. This could produce a large and unpredictable bump during transfer back to automatic, which must be avoided. To prevent windup, the manual output could be fed into the controller as the constant of integration, either at the input or the output of the "reset" (integral) restrictor. If fed into the output of the restrictor, the controller would respond immediately to any changes in the manual output. Unfortunately, however, this placed a switch downstream of the restrictor, where even a very slight leak could cause a loss in pressure resulting in an offset. As a result, the manual output was connected upstream of the restrictor, in which case the constant of integration lagged the manual output by the integral time constant.

While loading the constant of integration with the manual output prevented windup, the controller and manual regulator outputs were only equal if there were no deviation and if the regulator had not been adjusted recently (within the integral time). To avoid bumping while transferring in either direction, a three- or four-position switch was provided in some controllers such as the Foxboro models 52 and 58 pneumatic controllers and models 61 and 62 electronic controllers. When proceeding from the automatic to the manual mode, the switch was placed in a "seal" position, where the controller output was held at its last value, and the automatic and manual outputs were compared on a deviation meter. The operator would then adjust the manual regulator to balance them before transferring to the manual position. A similar procedure was followed in the reverse direction, with the operator adjusting the set point to achieve a balance before transferring to automatic. In this way, bumping could be avoided in both directions with a little care, providing that the transfer switch performed smoothly and did not leak. However, the act of moving the set point to produce a balance prior to transfer could mean that the set point was not necessarily at its desired value, and repositioning it after transfer could still cause a bump.

Achieving true bumpless transfer

Bumpless transfer to manual is a simple task if the controller's integrator also doubles as the manual regulator. This was the method used in the Foxboro model 62H electronic controller. Adjustment of the manual output was accomplished by pulsing the integrator in either direction at either of two speeds using a spring-return switch. The output would ramp in response to pressure on the switch, stopping in place when the pressure was released. Transfer between automatic and manual modes simply changed the sources of input to the same integrator and was therefore bumpless in both directions. This technique also allowed transfer to be effected easily from a remote source by a relay, opening new possibilities for process automation.

As an example of the possibilities, consider a wastewater treatment plant where the water to be treated is pumped into a neutralization vessel periodically as it accumulates in a holding facility. While the pump is running, wastewater flows at a constant rate and is neutralized by a pH controller manipulating sequenced acid and caustic valves. When a low-level switch in the holding basin stops the pump, the air supply to the two reagent valves should be shut off by a solenoid valve. At the same time, the pH controller should be transferred to manual to prevent windup while the loop is open. While in the manual mode, its output should be held at its last value while in the automatic mode, because when the pump restarts, the most likely requirement for reagent (i.e., the load) is the same as experienced prior to shutdown.

Velocity or incremental digital algorithms also use a common integrator to produce their output, allowing the same simplicity in effecting bumpless transfer, which was one of the reasons for their widespread use in the early digital systems. While this works well for single-loop control, it introduces complications in a cascade system. The integrator between the primary output and the secondary set point does not stop integrating when the valve limits. Special consideration must then be given to protect the primary integrator from windup when the secondary saturates.

Bumpless transfer from manual to automatic offers an opportunity to avoid proportional action on set-point changes, as long as they are introduced while the controller is in the manual mode. However, if proportional action is desired, then the change can be introduced while the controller is in the automatic mode. The model 62H controller had some other distinctions of its own. It was a noninteracting PID controller, but its unique circuitry made it difficult to prevent integral windup during saturation. In addition, *any* adjustment made to the proportional band while the controller was in the automatic mode bumped the output such that users were advised to transfer to the manual mode while tuning.

Another limitation of this controller was the absence of what is called *hard manual.* The automatic portion of the controller could not be removed or replaced without losing the manual output to the final actuator. In many applications—in power plants, for example—this condition was considered unsafe, since loss of a control signal could bring the plant down. To achieve bumpless transfer between automatic output and hard manual required a method of automatically balancing the manual output against the automatic output prior to transfer. Moore Products Company solved the problem by driving the manual and set-point regulators in their pneumatic controller with a pneumatic turbine through a feedback loop. The manual regulator was then forced to follow the automatic output while the controller was in automatic so that transfer would be bumpless. Yet the output of the regulator was determined by shaft position and was therefore "nonvolatile" and could be repositioned by hand. Similarly, the set-point regulator was driven to keep the outputs balanced while in the manual mode. Again, this left the set point as a variable that had to be repositioned to its desired value once automatic operation had resumed. This feature is called *set-point tracking,* which has an advantage in some applications such as in secondary loops of cascade systems and is discussed in that context later.

The Foxboro series 130 pneumatic and SPEC-200 electronic controllers had separate manual stations and forced the currently unused output to follow the output in use by means of amplifiers. In the automatic mode, the output m relates to the set point r, controlled variable c, and output bias b as

$$m = b \pm \frac{100}{P}\left(r - c - D\frac{dc}{dt}\right) \tag{9.1}$$

where the sign is determined by the action selected for the controller; P is the proportional band, and D is the derivative time. Bias b represents the constant of integration, whether the integration is performed on the deviation itself or by positive feedback of the output through a lag. When the controller is placed in manual, b must be manipulated to keep the automatic output matched to the manual output. This can be done either through a feedback loop or by a back-calculation. For the two controllers mentioned above, a high-gain feedback loop matched the two outputs by manipulating b directly, i.e., without passing through the integral time constant. With digital controllers, it is more expedient to back-calculate b based on current values of output, set point, and controlled variable:

$$b = m \mp \frac{100}{P}\left(r - c - D\frac{dc}{dt}\right) \tag{9.2}$$

where m now represents the manual output.

Bumpless transfer with proportional control

Use of either of the preceding balancing methods creates a problem for the proportional controller. In order to satisfy the requirements of the few applications that require proportional or PD control as described in Chap. 3, the output bias *must* remain constant. Attempts to use a PI or PID controller in these applications will fail because b is readjusted whenever the controller is placed in manual and, for some controllers, whenever the output reaches a limit. There is then no way of predicting its final value when the controller is returned to automatic or the output returns from its limit.

Bumpless transfer can still be achieved with proportional and PD control, however, by using two different values of b: a temporary value used to balance the outputs when in manual and a permanent value that is approached in the steady state. After the bumpless transfer, the bias returns to its permanent value on an exponential decay whose time constant is adjustable. This behavior is relatively easy to achieve in a PI or PID controller where integration is achieved by feedback of the output through a first-order lag; the permanent bias is placed at the input to the lag, and the back-calculation is applied at its output. While in the manual mode, the internal feedback loop or back-calculation would be adjusting the output of that lag to provide the necessary balancing. Following transfer, the bias would then be free to approach the input to the lag, its permanent value. In these controllers, the integral time constant may be renamed *balance time*.

Power interruptions

Always an issue in controlling process plants is the action a controller is to take when power is interrupted and again when it is restored. Final actuators should be selected to fail safely upon loss of signal or power. With this in mind, I designed the controls for a high-energy fuels plant in the 1950s with a "panic" switch on the control panel that vented the air supply to all controllers. It was used at least once by an operator to shut down the plant when a fire broke out. The control signals to the valves were carried by polyethylene tubing for the same reason: A fire burning through the lines would cause the valves to assume a safe position, shutting off fuels and chemicals while maximizing coolant flows.

If the air supply failed, however, the automatic or manual mode of the controller did not change. It was usually necessary for the operator to place all the controllers in the manual mode prior to restarting the process if the outage was long enough to change the state of the plant significantly. However, windup was not a problem on loss of air supply in that the high limit of the controller and of the constant of integration was the air-supply pressure. The simplicity of pneumatic controllers has some important benefits. Electronic analog controllers share many of the same fail-safe features, with added functionality, but the digital world is quite different. The constant of integration in a digital controller is a number independent of the supply voltage. As a result, a 16-bit digital controller can wind up to $\pm 32{,}767$ (which is $\pm 2^{15}$) if no protection or limit is provided.

It is also possible to configure a digital controller to assume any desired state upon the application of power. This applies to its auto-manual status, its set point and output, and even its PID settings. It is therefore incumbent on the system designer to make the selections appropriate to the status desired on a power restart. A question remains, however, on the length of power outage that would require these special restart conditions. Interruptions of less than a second are probably the most common, and these are too short to shut down a typical process plant. Therefore, the controls should be programmed to restart in their existing status on a short outage and in the specially designated status on longer outages. The decision on the duration that distinguishes the two types of outages then must be made. In some plants, controls are equipped with uninterruptible power sources, permitting continued operation on standby power for hours after an outage, in which case the restart mode may become a moot point.

Output Limits

Manipulated variables are always limited at both an upper and lower bound, and the controller should reflect the same limits. The earliest controllers had no output limits and, as a result, could wind up to the full supply pressure (125 to 150 percent of scale) or atmospheric pressure (-25 percent) depending on the direction of the deviation. Adding output limits to the controller does not by itself prevent windup, but it does limit the amount of windup possible. Prevention of windup is discussed at length later in this chapter. At this point, the role of output limiting is presented in conjunction with the design of the controller.

Position limits

The upper and lower limits on the position of the manipulated variable are nominally 0 and 100 percent, and these same limits should be set into the controller output to signify both to the controller and to the operator that the loop is open when they are reached. However, 0 and 100 percent are only *nominal*. Some pneumatic valves open at 4 to 5 lb/in^2 (8 to 16 percent of scale) as required to close completely against line pressure. Also, some manipulated variables require very specific limits, and these applications deserve special consideration. For example, the lowest allowable fuel flow to a burner may be set at 20 percent of scale to avoid flameout. It is customary to fix a mechanical stop on the fuel valve to prevent an accidental flameout rather than to rely on the low-limit setting of the controller, which is more subject to inadvertent or unauthorized adjustment. In these cases, the controller's low limit should be set equal to that of the valve, and windup protection should begin there. Additionally, the observation that the controller output cannot be decreased below that point will notify the operator that it is useless to attempt to control beyond it, either manually or automatically. These output limits are normally placed downstream of the auto-manual transfer switch and upstream of the integral feedback path, as shown in Fig. 9.1. In this way, the limits apply equally to the manual output and the constant of integration. The latter does not prevent windup by itself, but it keeps windup from exceeding the output limits.

In some situations, it may be desirable to allow the operator to override the limits when the controller is in the manual mode. Obviously, this would not be the case in the presence of a mechanical stop, since

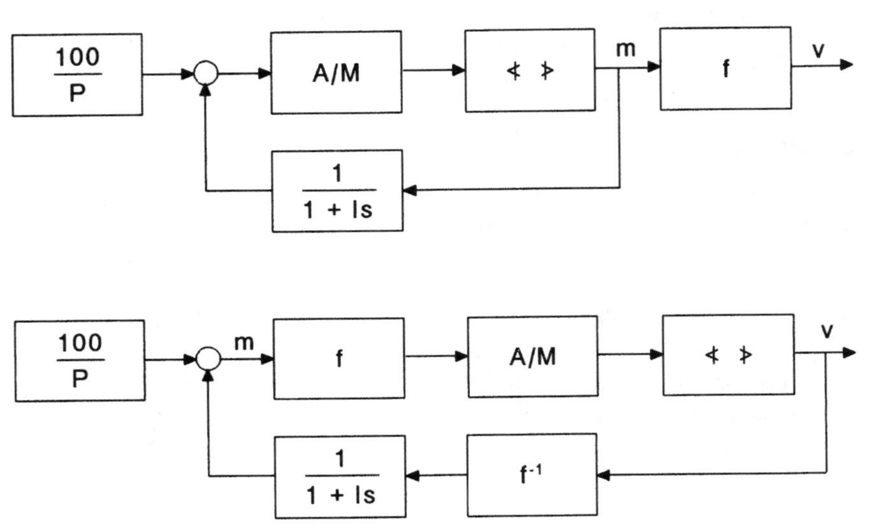

Figure 9.1 An analog controller is likely to have valve characterization added externally (*above*), whereas it can be made internal to a digital controller (*below*).

any signal override then would have no effect. There are some processes, however, that can only be controlled automatically over a limited load range without going unstable. This is true of most boilers and heat exchangers, because their deadtime, time constants, and steady-state gain all vary inversely with flow. Below about 25 percent load, control tends to deteriorate markedly unless suitable adaptation is used. Startup and low-load operation may have to be conducted manually or with another control system. In a case such as this, separate limits for automatic and manual control may be employed.

Output limits may even be set remotely, either changed among two or more values using logic, or varied numerically from a programmed source. A limit also may be driven from the output of a second controller, allowing the second controller to override the first. If the connection is made to the high limit, the resulting output becomes the lower of the two controller outputs—a low-signal selection; conversely, making the connection to the low limit produces a high-signal selection. The former is more common, in that it provides fail-safe action, a signal loss driving the selected output to zero. The first controller is protected from windup because its integral feedback signal is taken downstream from the limiter. The second could wind up, however, unless the limited output of the first controller is also connected to the integral feedback port of the second controller. This second connection is not internal and must be made intentionally.

It may be possible for the high and low limits to overlap, especially if they are set from different sources. In this case, the high limit should be given preference over the low limit, since driving the controller to the lower output, it would place the loop in a safer condition.

The output characterizer

Output characterization can be added to a controller in either of two ways, both shown in Fig. 9.1. In the upper configuration, the characterizer (f) is external to the controller, with auto-manual transfer, output limits, and integral feedback applied prior to characterization. This is the method that would be used if the characterization were conducted within the valve positioner, for example, or by an analog function generator connected to the controller output. The output signal from the controller and the limit settings would not reflect the actual position of the valve but rather the inverse function of its position. While this is the configuration used with analog controllers and may be the easiest to apply when adding characterization to an existing loop, it does not give the operator a handle on the actual position of the valve. In addition, the limits set in the controller may not match the valve's limits.

The second configuration in Fig. 9.1 is available in some digital controllers such as the Foxboro 760 series and is recommended as giving the operator more direct control over the valve. Characterizer (f) is placed upstream of the transfer switch and the limits; the output of the controller is then the true position of the valve. However, integral windup protection requires that the positive-feedback loop be connected downstream of the transfer switch and the limits. However, if the feedback signal and the output from the summing junction are not identical, the controlled deviation will not approach zero in the steady state. Therefore, placing function f in the forward path of the positive-feedback loop requires the inverse function f^{-1} in the feedback path. If the characterizer consists of a set of xy pairs connected by straight lines, then the inverse function is simply the same set of pairs with input and output reversed. (The function must be monotonic, of course, since a double-valued function could not be inverted.)

If the universal hyperbolic characterizer described in the preceding chapter were used in such a configuration, the valve position v would be produced by the function

$$v = f(m) = \frac{m}{L + (1 - L)m} \tag{9.3}$$

where L is the linearity adjustment which determines the degree of curvature to the function, a value of 1.0 producing a linear function. This calculation performed in the forward path must be matched by the inverse function in the feedback path:

$$f^{-1} = \frac{vL}{1 - (1 - L)v} \tag{9.4}$$

It may be risky to use this configuration with an analog controller, in that the feedback function may not be a perfect inverse of the forward function. However, a digital controller presents no problems in this respect, because both calculations can be made with absolute accuracy. Protection need only be provided to avoid such errors as "divide by zero," which in the preceding functions can be assured by placing a low limit on L to keep it above zero and to keep m from going negative.

Velocity limits

Most valves have a limited stroking speed, called a *velocity* or *rate limit*. It may be expressed as the maximum stroking speed in percent per second or inversely as the time required to travel full stroke. This is a nonlinear dynamic function, presenting a slower response to larger signals than to small ones.[1] Velocity limiting is not a destabilizing function like deadband, but it does present some response problems. If the controller is tuned for stability in the presence of a small deviation, the velocity limit may not interfere with the response at all. A large upset, however, will require stroking the valve over a large distance, which requires more time. This extends the period of oscillation longer than compatible with the integral setting of the controller. Observe the excessive overshoot shown in the first response in Fig. 9.2, caused when the ramping valve position cannot keep up with the sinusoidal controller output. The loop remains damped, fortunately, so that eventually the velocity of the controller output falls within the stroking speed of the valve—the period and decay ratio then return to normal. Detuning the integral mode can protect against this excessive overshoot for an expected size of disturbance, but this will retard recovery for smaller disturbances, increasing the integrated error by the increase in integral time.

A preferred solution to the problem requires using the actual stem position of the valve as feedback to the integral mode of the controller. This automatically retards integration when the valve is unable to keep up with the controller output, giving uniform overshoot and decay regardless of the disturbance size, as shown in the second response in Fig. 9.2. The first half cycle is as slow as before,

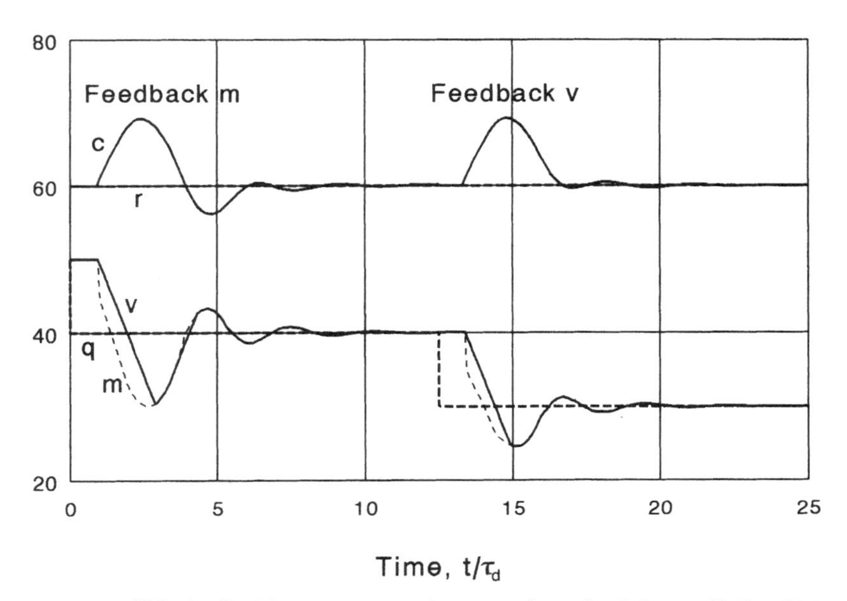

Figure 9.2 Velocity limiting causes overshoot on a large load change (*left*), which can be eliminated by taking integral feedback downstream of the limiter (*right*).

but this is unavoidable, given the velocity limit; IAE is reduced almost 20 percent by eliminating the overshoot.

This technique is effective for motorized valves where stem position is measured and used as feedback to a position loop, a mechanism common in power plants. For valves without such a position measurement, a velocity limit can be inserted in the controller's output path, where function f appears in the lower structure of Fig. 9.1. This signal is representative of the true stem position if the velocity limiter is matched to the valve motor or is slower. Taking integral feedback from the controller output will then have the same effect as using the actual stem position, provided there is no dead band in the valve. It might be argued that the same effect would be achieved if the velocity limiter were instead inserted where the f^{-1} block appears in the feedback path. However, it may be desirable in some installations to slow the valve below its inherent stroking speed, and the velocity limiter in the forward path allows this; there is then no need for an additional limiter in the feedback path. Most valves have the same stroking speed in both directions, but the velocity limiter in the controller could be set for two different speeds if the application required.

The Tracking Mode

Tracking involves forcing the controller to follow some remote signal to prepare it for automatic operation once a start command is given. The controller could be applied to a batch process having automatic startup and shutdown, or it could be a backup for a parallel controller, taking over upon a failure of the latter or on operator intervention. Either the output or the set point or both may be made to track remote signals. Two different inputs are actually required: the signal that is to be tracked and the command that starts and stops tracking.

Output tracking

A common requirement is the use of a field or local controller to back up a remote or central controller. It must be able to take over control bumplessly when a signal indicates a failure of the remote device. An operator also must be able to override the remote signal, transferring control to the local device after observing some problem in the local area. Transfer must be smooth and bumpless in both directions, which is only possible if both outputs are equal at the time of transfer. Therefore, the inactive controller must be forced to track the active controller continuously.

Output tracking is accomplished in the same way as manual control, except that the controller output is forced to follow the remote tracking signal instead of its own manual output. The internal bias is back-calculated as shown in Eq. 9.2 to keep the two outputs matched. When the tracking command is removed, automatic control begins from that last output. If any deviation exists, the controller output will integrate as needed to remove it rather than produce a proportional bump. When one controller is used to track another for backup purposes, both normally require the tracking feature to provide bumpless transfer in both directions.

Set-point tracking

Set-point tracking may or may not be required. If there is a safe position for the set point of the backup controller, which there may be for those intended to enforce safe operating limits, set-point tracking is not needed and may be detrimental. However, in cases where the set point of the remote controller is subject to frequent adjustment, the set point of the backup controller ordinarily is expected to follow it. This can be accomplished by retransmitting the set point from the remote controller to the backup controller or by connecting the controlled variable to the set point of the backup controller while it is tracking and then freezing it there when the tracking command is removed.

For secondary controllers in cascade systems, set-point tracking is usually desirable. The reason is that there is no fixed or predictable reference point that could be determined in advance, because its required value varies with load or other unmeasured influences. Set-point tracking in this case, however, is accomplished by forcing the *primary* controller output to track the *secondary* controlled variable. This keeps the deviation of the secondary controller at zero while the primary controller is tracking. In this application, the primary controller should be placed in the tracking mode whenever the secondary controller is in the manual mode. Upon transfer of the secondary controller to the automatic mode, the tracking command is removed from the primary controller so that it is also in the automatic mode.

An instance where secondary set-point tracking is undesirable is in the control of batch reactor temperature. Transfer of the secondary controller to automatic signals start-up, in which case full heat is to be applied immediately to the reactor. If tracking is used, however, there will be no proportional bump upon transfer, causing the primary controller output to *integrate* to its set point. This can prolong the heating cycle unnecessarily, delaying completion of a batch of product.

Integral Windup Protection

The cause and effect of integral windup were described briefly in Chap. 3. In essence, a sustained deviation caused by an open control loop will be integrated by the controller until the constant of integra-

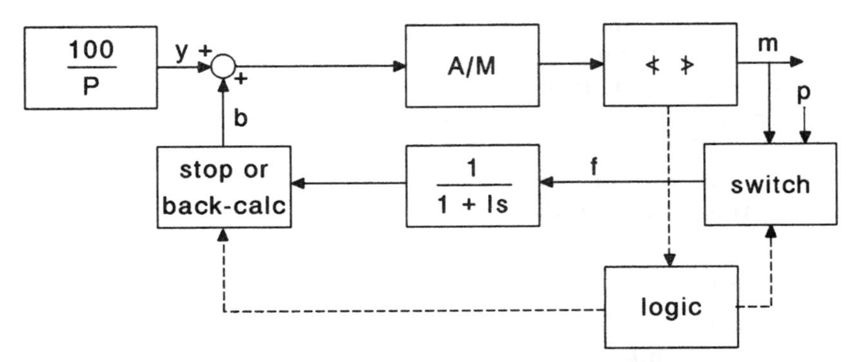

Figure 9.3 Windup protection in a single loop requires some type of logic to interrupt integration.

tion—effectively bias b above—reaches one of the output limits. With b at the limit, output m in Eq. 9.1 will not retreat from the limit until the combination of e and the derivative of c cross zero. In the absence of derivative action, the output will remain at its limit until e changes sign, making overshoot unavoidable. This is also true of single-stage controllers where derivative acts on the output rather than on the controlled variable. In Chap. 6, setting derivative time greater than integral time was found to be effective in avoiding set-point overshoot on lag-dominant processes, even with a woundup controller. Described below are methods that can avoid overshoot caused by windup without the use of derivative action.

Figure 9.3 shows logic added to an interacting PID controller to interfere with integration whenever the output of the controller reaches a limit. Some of these methods apply equally well to noninteracting controllers or when integration is accomplished directly on the deviation, but all of them apply to the positive-feedback loop shown in the figure.

Stop integration

The first method simply stops integration whenever the output reaches a limit, just as if the integral time of the controller were suddenly increased to infinity. This effectively locks b at the value it had when the limit was last reached and applies equally well whether integration is performed on the deviation or by output feedback. When integration stops, b will have reached the value given by Eq. 9.2, where output m has the value of whichever limit was encountered.

If the limit were encountered after a very slow integration of a very small deviation, b would have a value essentially equal to the limit, and the controller would therefore be wound up, just as if no windup protection were employed. In the case of a sustained deviation e caused by an open loop, the controller would integrate until a limit were reached; at this point, b would differ from the limit by $100e/P$, since there would be no derivative contribution. A third possibility is a sudden change in controller output reaching a limit, caused by a set-point change $\Delta r > P$ or derivative action on a sudden change in the controlled variable. This would leave b at the last value it had before the upset. In all these cases, b will fall between the high and low limits m_h and m_l but with an indeterminate value dependent on the deviation and the derivative of the controlled variable at the point when the limit was encountered.

The controller output will then leave the limit at whatever combination of conditions caused it to reach the limit. If the process being controlled were continuous and the limit were reached due to an excessive load, holding b at its last value is probably the best action to take. It assumes the process will return to the control range on the same path it left. Responses of simulated PI control loops are compared in Fig. 9.4. The load had been earlier stepped from 10 percent above the low limit to 10 percent below it. At the beginning of the plot in the figure, the load is stepped back. The controller with integration limited comes off its output limit as the controlled variable crosses the set point because b and m are equal at the limit; as a result, over-

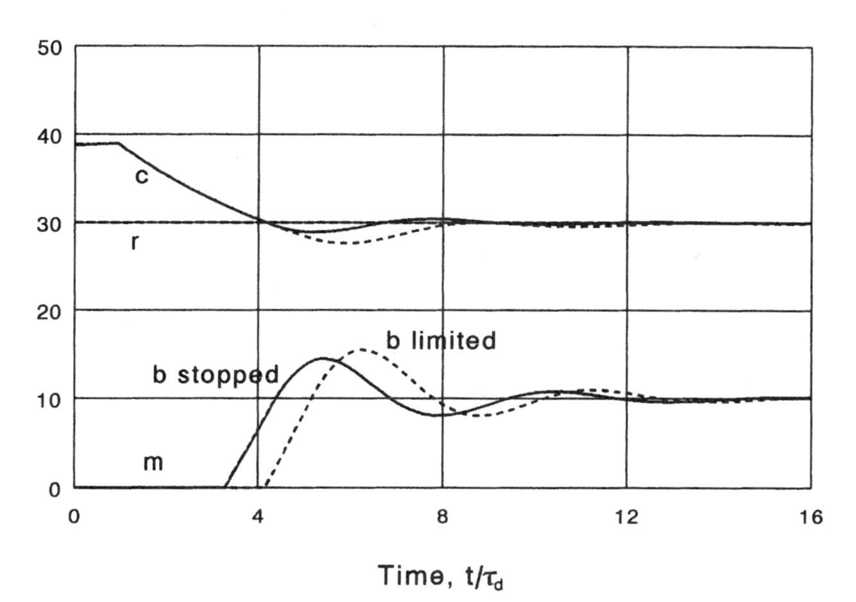

Figure 9.4 When the load returns from beyond a limit to a value it had before the limit was reached, stopping integration results in less overshoot than simply limiting integration.

shoot is unavoidable. The controller with integration stopped comes off its limit earlier because b is closer to the new load than it is to the limit, resulting in less overshoot.

With b limited, the PI controller caused an area overshoot of 29 percent, whereas with b stopped, the overshoot was reduced to 8 percent. The area overshoot was found by comparing the integrated absolute error with the absolute value of the integrated error for the response curves:

$$\Omega_A = \frac{\text{IAE} - |\text{IE}|}{\text{IAE} + |\text{IE}|} \tag{9.5}$$

which formula applies if the set point is crossed but once. The area overshoot will be remembered as a measure of set-point response from Chap. 6. With PID controllers tested under the same conditions, less overshoot is experienced for both methods: 2 percent with integration stopped compared with 12 percent with b limited.

Small set-point changes to PI and noninteracting PID controllers on a non-self-regulating process produce an area overshoot of 100 percent because they drive integrated error to zero. Filtering of the set point was found to be effective in preventing overshoot on set-point changes only as long as the controller did not saturate. Steps large enough to cause saturation actually reduce area overshoot somewhat by clipping the manipulated variable. Stopping integration upon saturation can reduce this overshoot even more but not eliminate it, resulting in an integrated error that is the product of the set-point change and the process deadtime.

Another consideration in windup protection is the possibility of noise rectification that would produce a steady-state offset. Even without windup protection, a noisy output operating near a limit will have some of its peaks cut off by the limit, preventing the integrated error from being zero. All windup protection methods increase the potential for offset by integrating at different rates for unlimited and limited noise spikes. Stopping integration when the output limits causes some ratcheting, resulting in an offset proportional to the noise level and the difference between the mean output and the limit. A simulated output having a 15 percent noise amplitude (3σ) and a mean value 10 percent from the limit produced a mean offset of 0.05 percent without windup protection and 0.27 percent when integration was stopped at the limit. If the offset produced by such rectification is objectionable, filtering will be required.

The batch switch

In the case of a batch process, the exit path leading to the limited condition and the return path from it may be entirely different, often representing the termination and start-up of different production runs. This also may be true in the case of large set-point changes that bring on a grade change or the next in a sequence of stages. The optimal value of b when the next grade or stage is reached may be quite different from its last value. Stopping integration when a limit is reached is usually not acceptable for these processes.

One method for windup protection of batch processes applies a switch to the input of the integral lag in Fig. 9.3. If the output is within limits, the switch connects the output to the lag so that integration proceeds normally; if the output is at a limit, however, an adjustable constant called the *preload* is connected there instead. The switch forms a negative-feedback loop through the integral lag and the summing junction. If the preload is set away from the limit, tripping of the batch switch sends the output moving away from the limit at a rate proportional to the difference between b and preload p. The resulting change in output then returns the switch to the output signal, allowing the controller to integrate back to the limit, where the switch trips again.

In pneumatic controllers prior to about 1970, the batch switch was a three-way valve driven by a diaphragm connected to the controller output. It had perhaps 2 percent dead band and consequently caused the output to cycle with an amplitude of about 2 percent when near the limit. In later pneumatic controllers, a throttling relay was used, and cycling was thereby avoided, as it was in electronic analog controllers. Using the batch switch with a digital controller returns the possibility of cycling at a period of two sample intervals; the amplitude of cycling increases with the ratio of sample interval to integral time and with derivative action.

If the required bias as estimated from Eq. 9.2, using the limited value of m, falls between the limit and the preload, then cycling or throttling will continue, with b assuming a mean value equal to that result. Any subsequent change in deviation or the derivative of the controlled variable while the controller is limited will result in a new value of b, in contrast to the last method, where b was held at its last value when the limit was reached. Should the deviation continue to increase, b will move farther from the limit, continuing to satisfy Eq. 9.2. Any subsequent *reduction* in deviation will then cause the output to move off the limit *immediately*. This action is compared against stopping integration following an encounter with the low-output limit in Fig. 9.5. The output leaves the limit much earlier than for the controller with integration stopped, and as a result, the controlled variable undershoots the set point. IAE is 30 percent higher using the batch switch.

Yet there are applications where overshoot *must* be avoided coming out of saturation, and the batch switch does avoid it. A prominent example is the *antisurge* controller used to protect centrifugal and

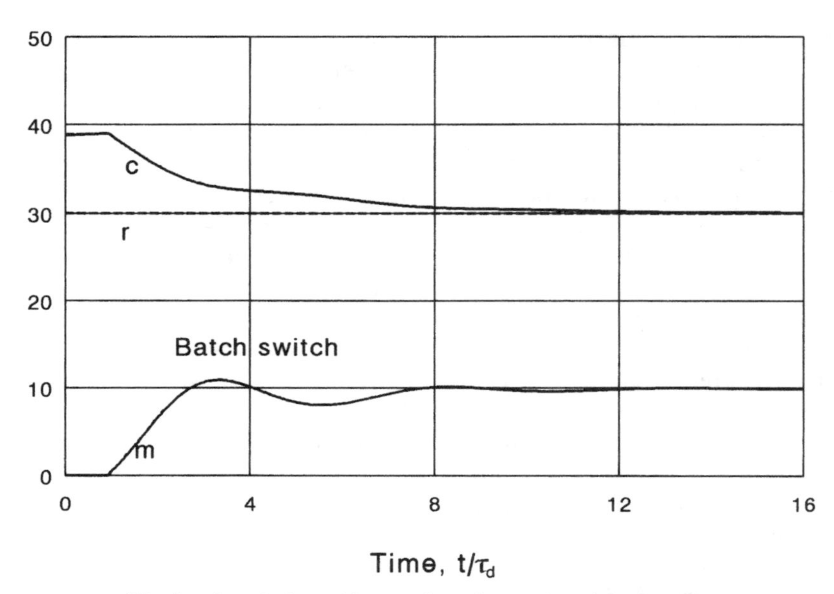

Figure 9.5 The batch switch avoids overshoot by moving the controller output as soon as deviation decreases.

axial compressors from a destructive oscillation known as *surge*. When the velocity of gas approaching the blades of a compressor falls below the speed of those blades, a stall develops. A negative-resistance range is entered wherein the compressor is unstable, causing flow to oscillate, actually reversing direction at intervals of about 1 second, stressing the blades. To prevent the compressor from operating in the surge region, a suction flow controller is used to maintain a minimum flow passing through it by recycling some of the compressed gas. Normally, the load is well above the set point of the controller, which responds by keeping the recycle valve closed. However, if the controller is in a wound up condition, it will not start to open the valve until the flow falls below set point, which could cause a temporary surge.

Avoiding windup by stopping integration may not be satisfactory in this application, in that the value of b at any point in time is indeterminate, and the controller could be operating at its limit for days or weeks at a time. If b were to approach the limit, no overshoot protection would be provided. A batch switch does provide this protection, however, by positioning b wherever necessary to keep the controller output just at the limit. Any reduction in flow will open the valve immediately, no matter how large the deviation, so that overshoot is always avoided. This is then the recommended solution for antisurge controllers.

Noise rectification is more severe using a batch switch than for stopping integration. Under the same noisy conditions used before to test for offset, the controller with a batch switch produced 0.8 percent offset—about three times as much as experienced when integration was stopped. The reason for the increase is that the batch switch drives b up and down at two different speeds. It is driven toward the limit at a rate proportional to $m - b$ and away from the limit at a rate proportional to $b - p$, where p is the preload setting.

A typical *batch* process presents intervals of open-loop conditions to controllers. For example, a batch reactor is emptied of product when the reaction is complete and the product has been cooled. The vessel must then be rinsed, evacuated, pressure tested, and refilled before reheating. While the vessel is empty, the temperature loop is open, although the controller may have been left in the automatic mode. Even if the next start-up were to begin with the controller in the manual mode, heating the batch from an ambient temperature of typically 20°C to a reaction temperature of perhaps 70°C will cause saturation because this deviation will typically be two to three times as great as the proportional band. It is a mistake to use a controller having no proportional response to set point or to introduce the set-point change while the controller is in the manual mode, because then the output will ramp upscale rather than step, delaying the approach to set point. The shortest time to set point will be attained when the controller saturates immediately.

Saturation begets windup unless protection is applied. If the heating valve remains fully open until the reactor temperature crosses set point, as it would with a woundup controller, the resulting overshoot would spoil the product. Figure 9.6 describes the simulated response of a batch reactor to said massive set-point change using a PI controller with a batch switch. If preload p is set at the upper limit of 100 percent, the controller is essentially woundup, since that is the value assumed by b. This results in a large overshoot.

The preload has no effect for deviations smaller than

$$|e| < \frac{P}{100}|m - p| \tag{9.6}$$

where m represents the output at the limit. In the case of the antisurge flow controller, the proportional band is typically 200 percent—much larger than e is ever likely to be. In this case, p can be left at the opposite limit, and b will never approach that value. In the case of temperature control, however, setting p at the opposite limit would make $m - p$ in Eq. 9.6 100 percent at most, and e can easily exceed the proportional band P. In this situation, the batch switch remains in the preload position, so b assumes the preload value as long as the output is limited.

The second pair of curves in Fig. 9.6 shows the effect of preload being set at 30 percent relative to a process load q of 50 percent. The

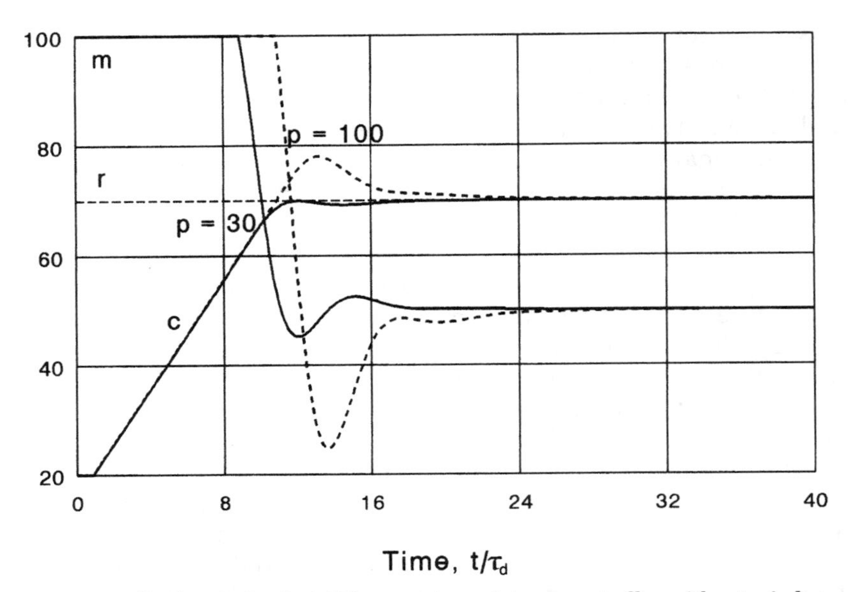

Figure 9.6 Setting preload at 100 percent emulates a controller without windup protection; a value of 30 percent avoids overshoot in the presence of a 50 percent load.

controlled variable responds by just touching the set point before bouncing back slightly and then settling in. If the preload were set precisely equal to the load, which represents the steady-state value of controller output, some overshoot would result; integration begins when the output leaves the limit, which occurs *before* set point is reached. The controller then integrates some error and increases b from its preloaded value, but b must equal load q in the steady state, requiring that error to be integrated back out again.

Preload also can be set too low, resulting in undershoot. The value of 30 percent used in Fig. 9.6 produces a response curve that is optimal with respect to IAE. Therefore, its setting must be optimized relative to the anticipated load, which is subject to change with set point and recipe and even with the condition of the heat-transfer surface. The proportional and integral settings used to produce the response curves in Fig. 9.6 are about 20 percent higher than optimal for load rejection. This is necessary to minimize the bounce that follows the first encounter with the set point.

Most temperature controllers will have derivative action, which does affect the shape of the response curve. If derivative acts on the controlled variable, it remains active while the controller is saturated, causing the output to leave the limit sooner. This rounds the response curve noticeably, actually prolonging the settling time.[2] Better results are likely to be achieved by setting derivative time greater than integral time, as recommended for set-point changes in Fig. 6.11, which does not require any logic and which works equally well whether the controller is saturated or not (but applies only to interacting PID controllers).

Single-stage PID controllers, i.e., those having derivative control acting on the output, behave quite differently. When saturated, derivative control is disabled and then becomes enabled abruptly when the limit is left. The batch switch is quite effective with these controllers, and for many years they have been used successfully to control batch reactors.

Back-calculation

Still another method for windup protection involves back-calculating b based on Eq. 9.2 using a value of m representing the output limit. This method has exactly the same steady-state result as the batch switch but is imposed immediately instead of through the integral time constant. It has the advantage of eliminating the feedback loop produced by the batch switch and its consequent cycling. A limit of p can be placed on the back-calculated value of b, again as done with the batch switch. Otherwise unrealistic values of b, negative or exceeding 100 percent, could result whenever deviation increased beyond the proportional band. Recovery from a limited condition with back-calculation will duplicate that observed using the batch switch.

There is one substantial difference, however, relative to its sensitivity to noise rectification. Under the same noisy conditions used to evaluate the other controllers, the offset produced using back-calculation was over 2 percent, almost three times that observed for the batch switch. The reason for the increased sensitivity is that each time the limit is touched, a new value of b is back-calculated *immediately,* rather than approached exponentially through the integral time constant. This gives a true ratcheting effect to the controller output, producing the larger offset. Controllers using this method of windup protection will require filtering of any noise present to minimize this problem. The antisurge flow measurement tends to be noisy, and filtering is not advised for its controller lest it interfere with speed of operation. However, any offset that should develop will lie on the safe (excess flow) side of set point, in that noise rectification tends to process b farther from the limit than it would be in the noise-free case.

Multiloop Systems

Any production facility will have multiple control loops, with interactions among some of the variables. To protect the most critical variables from upsets, various multiloop strategies are found helpful. Disturbances arising in the path of the manipulated variable can be suppressed using cascade control of a secondary variable that responds faster to the upset than does the primary controlled variable. Disturbances entering from outside sources are countered by feedforward control, and those arising from loop interaction are rejected by decouplers, although the two methods are essentially identical. Finally, the case where there are more controlled than manipulated variables is examined. Each example poses potential windup problems not encountered with single loops, and each has its own solutions to the problems.

Cascade control

A cascade system has a primary controller setting the secondary set point; only the secondary output connects to a final actuator and therefore is likely to encounter a limit. The primary output could be limited as well to avoid unsafe operating regions, but the difficult case has the secondary controller at a limit while the primary is not. Any of the preceding methods can be applied to protect the secondary controller from windup when it reaches a limit, but the primary loop is also opened at the same time. To protect the primary controller from windup requires a signal from the secondary loop indicating its limited condition.

A common approach to the problem is to pass the logic signal used by the secondary controller to stop its integration on to the primary controller as well, stopping integration there. While this method is reasonably effective, the cascade configuration offers an opportunity for improved dynamic response and robustness along with windup protection. Figure 9.7 shows the recommended connections.

Observe that the secondary controlled variable is sent to the primary controller as integral feedback. As long as the secondary controller is able to keep its deviation near zero, the primary output (secondary set point) and the feedback signal will approach each other, allowing primary integration to proceed toward zero deviation. Should the secondary loop open, however, due either to its controller being trans-

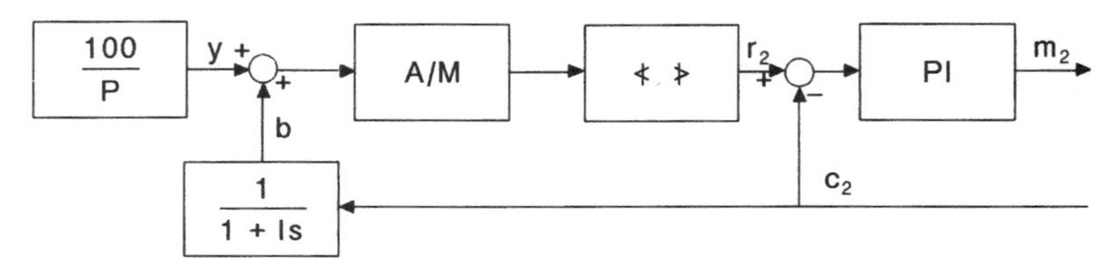

Figure 9.7 Using the secondary controlled variable as primary integral feedback provides windup protection and robust response as well.

ferred to manual or to a limit having been encountered, the primary integral feedback loop also opens. In this event, both controllers will approach steady-state offsets, related as follows:

$$e_2 = \pm\frac{100}{P_1}e_1 \tag{9.7}$$

Primary offset e_1 is determined by the limited or manual position of the final actuator, which also determines the steady-state value of secondary controlled variable c_2. The primary controller will then position its output, which is secondary set point r_2, to produce secondary deviation e_2. In another configuration, secondary offset would have no importance. With this configuration, however, Eq. 9.7 reveals that secondary offset produces primary offset. Therefore, integral action *must* be used in the secondary controller.

In some cascade systems, the primary output is made to track the secondary controlled variable when the secondary controller is in the manual mode. There is then no bump upon transfer to the automatic mode. With the configuration of Fig. 9.7, the primary controller can be left in *automatic* while the secondary is in manual. Transfer of the secondary controller back to the automatic mode may then produce a bump if there is a primary deviation, because Eq. 9.7 shows there also will be a secondary deviation. While the bump can be avoided by forcing the primary controller to track c_2, it is actually helpful in diminishing the primary deviation as quickly as possible, and tracking is therefore not recommended.

In Chap. 5, a model-based controller was created by inserting a deadtime block in the integral feedback path of a PID controller, thereby improving performance. In the configuration of Fig. 9.7, the *entire secondary loop* including the process is inserted there. This allows considerable reduction in the primary integral time until it is lower than that of the secondary controller, possibly even approaching zero. Performance definitely improves, and yet the system becomes more robust rather than less. As described at the end of Chap. 7, variations in dynamics of the secondary loop effectively vary the integral time of the primary controller in a compensating manner.

The effectiveness of this method of windup protection has been borne out experimentally in controlling batch reactors. Consider the reactor with its controls appearing in Fig. 7.9. Heating and cooling valves are shown operating in split range, first to raise the temperature to the operating point and then to remove the heat produced by the reaction. In a production reactor polymerizing vinyl chloride, it was customary to raise the contents of the reactor to the set point before applying any initiator. When the temperature had settled at set point, initiator was then injected by the operator, and the heating valve automatically blocked in to prevent applying heat to an exothermic reaction.

A control system applied by myself did *not* have the feedback connection described in Fig. 9.7. After the reaction had started, rising reactor temperature caused the cooling valve to open as shown in the simulated response of Fig. 9.8*a*. Without windup protection, the temperature overshot set point, and the cooling valve closed again. A few minutes later, the same response was repeated, and again and again, developing a limit cycle. Efforts made to retune both controllers had little effect, and the cycle continued until the end of the run. Although its amplitude was only about 0.5°C in the primary temperature, it had an adverse effect on product quality.

The problem was caused by the blocked heating valve opening the secondary loop whenever the cooling valve was closed. The problem was solved by using secondary temperature as primary integral feedback. Closing the cooling valve no longer caused windup, eliminating the source of the cycle. Figure 9.8*b* shows a smooth recovery following the injection of initiator and following later injections as well.

Adding feedforward control

Feedforward control is the conversion of a measurement of the disturbing variable into equivalent positioning of the manipulated variable to cancel the influence of that disturbance on the controlled variable. The calculations performed in the feedforward path are usually material or energy balances, manipulating flow rates in response to measured flow rates. Therefore, most feedforward systems calculate set points for a flow controller rather than positioning a valve, thus preventing the nonlinear characteristics of the valve from affecting the accuracy of the manipulation. Occasionally, feedforward control is used when the primary controlled variable cannot be measured on-line, and therefore, feedback control cannot be applied. More often, however, feedback control of the primary variable is combined with feedforward to set a secondary controller in cascade, as shown in Fig. 9.9. The relationship between the measured load and the manipulated variable is usually that of a flow ratio k with appropriate dynamic compensation represented by $f(t)$.

If the ratio of the manipulated variable to the load is fixed, that fixed ratio is applied, with the calculation biased as necessary by the primary controller to drive its deviation to zero. An example is the "three-element drum-level control" of boilers. Steam flow is the load variable q which sets feedwater flow set point r_2 in a 1:1 ratio, with feedback trim m_1 from the drum level controller:

$$r_2 = kq(t) + m_1 - 50 \tag{9.8}$$

where k would be 1, $q(t)$ is the dynamically compensated load measurement, and primary output m_1 is expressed in percentage of full

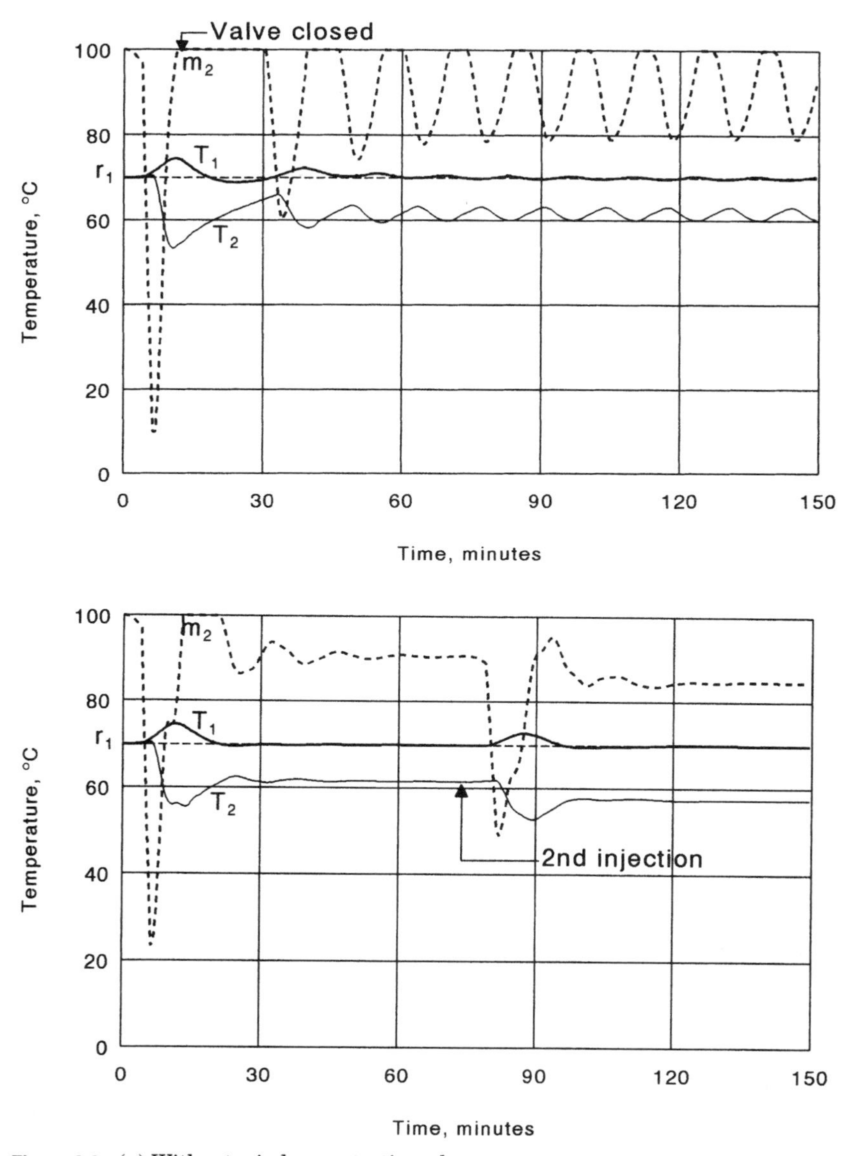

Figure 9.8 (*a*) Without windup protection, the reactor temperature controller cycles off the limit of the cooling valve. (*b*) Using secondary temperature as integral feedback to the primary controller eliminates the cycling caused by windup and adds robustness to the loop.

scale. The fixed bias of 50 percent establishes this as the normal output of the primary controller, i.e., when load and manipulated flow are perfectly matched. Windup protection for the primary controller requires this calculation to be reversed to produce the integral feed-

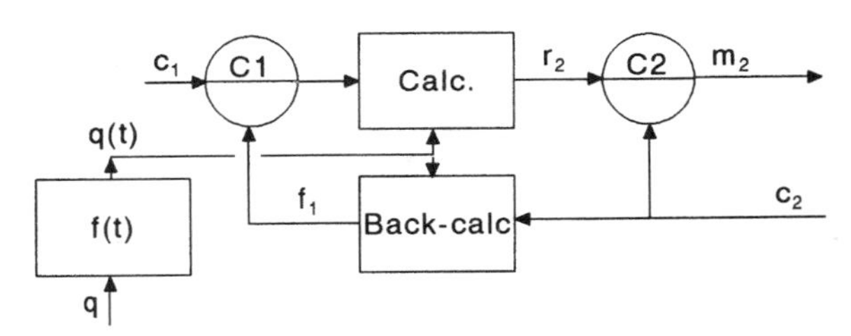

Figure 9.9 Any calculation inserted between the primary and secondary controllers requires a back-calculation in the integral feedback path.

back signal. To do this, c_2 is substituted for r_2, and m_1 is replaced with the feedback signal f_1:

$$f_1 = c_2 - kq(t) + 50 \tag{9.9}$$

In this way, f_1 matches m_1 as long as there is no secondary deviation, so primary integration proceeds normally.

In many feedforward systems, the ratio between the manipulated flow and the load is a variable, subject to unmeasurable influences. Then the primary controller should be applied to adjust the *ratio* as required to drive its deviation to zero, thereby recalibrating the feedforward calculation to the prevailing unmeasured influence. In this case, the feedforward and feedback calculations proceed as follows:

$$r_2 = kq(t)m_1/50 \tag{9.10}$$

where k now represents the normal or midrange ratio of the manipulated flow to the load. Here again, m_1 is expressed in percentage, having a normal value of 50 as established by that number in the calculation. The back-calculation solves for f_1 in terms of c_2 as before:

$$f_1 = 50c_2/kq(t) \tag{9.11}$$

Care should be taken when using multipliers and dividers in control systems to avoid multiplying and dividing by zero in the event of a loss in load. In the forward calculation, setting q to zero opens the feedback loop, preventing the feedback controller from having any effect on the controlled variable. A low limit on that input to the calculation should be set to represent the lowest load likely to be encountered while still maintaining control of the primary variable. The same limit needs to be applied to the load input in the back-calculation to ensure that integration proceeds normally while the feedforward calculation is limited.

Some engineers use the nominal 50 percent bias as a reference point to estimate how much effort is being required by the feedback controller to maintain zero deviation. If its output m_1 remains close to 50 percent under most plant conditions, then the feedforward loop is carrying most of the load. This is obviously desirable, in that the integrated error of the controlled variable is directly proportional to the change in m_1 between steady states. A steady-state value away from 50 percent then indicates either a miscalibration of the feedforward loop or the influence of an unmeasured load. The "health" of the system may then be inferred by observing m_1.

It is possible to incorporate the feedforward calculations within the primary feedback controller. This has the advantage of applying the output limits of the controller and its auto-manual transfer directly to the manipulated variable rather than only to the feedback trim signal as done in Fig. 9.9. This is commonly done in power plant controls where both feedforward and feedback loops are closed or neither is. It should *not* be applied to product quality control using analyzers that often must be taken off-line for recalibration and maintenance. Unavailability of the primary measurement would require transfer of the controller to manual, thereby losing the feedforward loop as well. There is also the disadvantage of being unable to observe the feedback trim m_1 independently of the feedforward contribution to the output.

Figure 9.10 shows two methods for combining feedforward and feedback within the primary controller. The dynamically compensated load $q(t)$ may be applied to forward and back calculations, as done in Fig. 9.9, but upstream of the auto-manual transfer station and the output limits. Both summation and multiplication should be provided in the forward path, and subtraction and division in the feedback path selected, as fits the application. Because all signals are contained within the controller, there is really no need for the 50 percent scaling factor appearing in Eqs. 9.7 to 9.11.

An alternative method used quite extensively has the load signal summed with the integral feedback signal after passing through the differentiator shown as the lowest block in Fig. 9.10. Differentiation is required so that no steady-state correction is applied to the integral feedback signal, since that would cause offset between the primary controlled variable and its set point. This allows a single calculation to be made; ordinarily, it is a summation—multiplication would offer no advantage since the steady-state gain of the multiplier would have to be 1.0.

The time constant F for the derivative term applied to the load input is adjustable, as required to set the steady-state gain of the output with respect to the load:

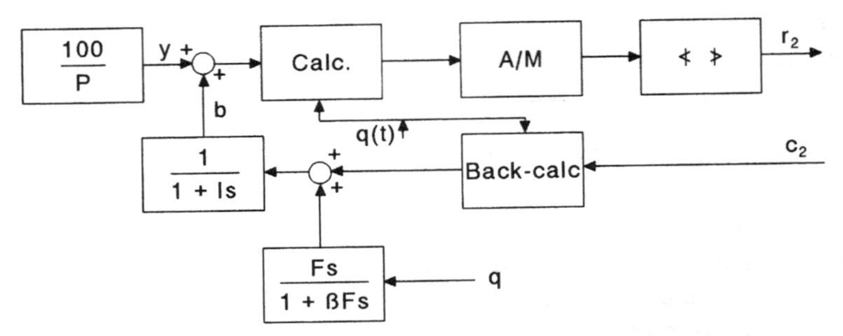

Figure 9.10 Feedforward compensation may be incorporated within the primary feedback controller either by applying a forward and back calculation or by adding a load derivative to the integral feedback signal.

$$dr_2/dq = F/I \tag{9.12}$$

This dependence of forward-loop gain on integral time I poses two problems. With analog controllers, time constants cannot be set with anywhere near the accuracy required for feedforward control, ± 20 percent error being a typical specification. Second, whenever the integral time of the feedback controller is retuned, the feedforward-loop gain changes as well. Consequently, this has been the least effective method of applying feedforward control. In a digital controller, however, both these objections may be avoided, with time constant F set in an accurate and fixed ratio to integral time I.

The differentiator also must include a filter, shown in Fig. 9.10 having a time constant of βF. This filter acts as a lag compensator in the forward loop, which is often required. Any additional dynamic compensation would have to be imposed on the load measurement upstream of the differentiator.

Decoupling systems

When two or more control loops interact through the process, a common solution is to apply decoupling. This technique is essentially identical to feedforward control, in that the output of one controller acts as a load on the other loop. Here again, we are concerned with preventing windup of feedback controllers having calculations interposed between their outputs and the manipulated variables.

A common example of decoupling applied to the top of a distillation column is shown in Fig. 9.11. Overhead vapor is condensed and accumulated in the reflux drum, whose liquid level must be controlled. The liquid is split into distillate product and reflux that is returned to the column. Column temperature, which is an indication of product composition, is a function of the ratio of these streams, while liquid level is a function of their sum.

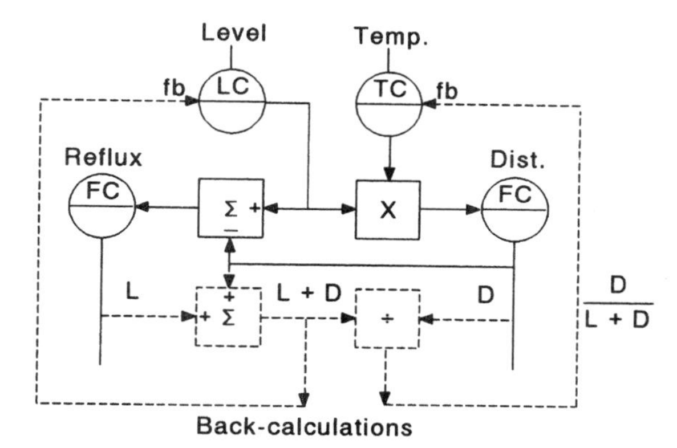

Figure 9.11 Liquid-level and temperature loops are decoupled by manipulating total liquid flow and flow ratio, respectively.

Ryskamp[3] devised this scheme to decouple the level and temperature loops; the level controller (LC) manipulates reflux plus distillate, while the temperature controller (TC) manipulates the ratio of distillate to the sum, which is the LC output. In the figure, the two controller outputs are multiplied to set distillate flow, while reflux flow is set as the difference between the LC output and measured distillate flow. The system regulates the column quite stably, responding equally well to upsets in the heat balance and in the column feed. However, it tends to be difficult for operators to transfer from the manual to the automatic mode, especially if they are more accustomed to other configurations. For example, it might be possible to place both the TC and LC in the automatic mode while leaving the reflux flow controller in the manual mode. Both primary controllers would then be competing for a single manipulated flow, and both could wind up.

To eliminate windup problems and facilitate transfer, back-calculations have been added to the loops as shown in Fig. 9.11. The feedback signal to the LC is the sum of the measured flows. If the reflux flow controller is in the manual mode or its output is limited, the sum of the measured flows will not equal the LC output, and windup will be averted. Similarly, the feedback signal to the TC is the ratio of measured distillate flow to the sum of the measured flows. Failure of the distillate flow controller (FC) to respond to the TC output will be indicated by no change in this feedback signal, again avoiding windup.

Having established the two flows, either with the FCs in manual or locally set, the LC may be transferred to the automatic mode. Switching the reflux FC to remote set in the automatic mode may bump its set point if there is a level deviation, but level control will be quickly established. The prevailing flow ratio will be properly conditioning the TC to take control of temperature whenever the distillate flow loop is closed. As before, a proportional bump may result if the loop is closed with the TC already in automatic, or it may be avoided by having the TC output track the feedback signal before closure of the distillate flow loop.

Selective control loops

The output limits described earlier in this chapter are considered *hard* constraints, in that they can be rigidly enforced by either mechanical stops or limiters of various kinds applied to manipulated variables. Constraints also may have to be applied to some controlled variables, but these fall under the category of *soft* constraints, in that a controller is required to enforce them. The effectiveness of this enforcement is limited by the same factors that limit all control loops: best practical control and the selection and tuning of the controller. Where feedforward can be used, the limiting factors are the deadtime difference between the manipulated and load paths and the accuracy of the match between the feedforward model and the process.

One example of a constrained controlled variable was the protection against compressor surge described relevant to the use of the batch switch. Suction flow was allowed to increase above set point, but the antisurge controller tried to keep it from falling below on a decrease in load. The batch switch biased the controller to open the recycle valve before set point was crossed to honor that constraint upon leaving a limited condition. However, while controlling at set point, the flow loop is subject to the same disturbances as other loops, which can be of unpredictable severity and in either direction. The controller cannot guarantee that flow will not fall below set point, and therefore, the set point should be positioned above the surge region by a margin likely to be sufficient for all foreseeable disturbances. Yet such a margin ought to be as narrow as possible, since excess flow through the recycle valve wastes the power used to compress it.

Antisurge control is an example of a single loop consisting of a controller and its dedicated valve, yet there are many processes requiring constraints on more controlled variables than there are valves. In this case, the controllers must be shared among the valves on the basis of need, i.e., proximity to constraints. A common example in distillation is the constraint placed on column differential pressure to avoid flooding the trays. As long as differential pressure is below the set point of the dPC in Fig. 9.12, the temperature controller is free to manipulate steam flow; this is considered normal or unconstrained operation, since temperature control provides regulation over product quality. However, if too much steam is applied to the reboiler, as required by an excessive

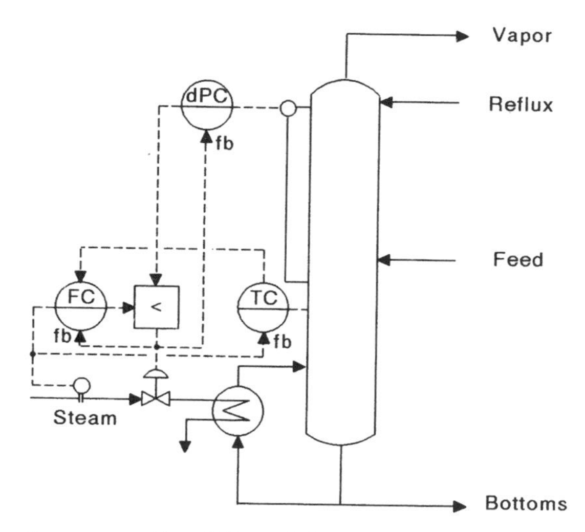

Figure 9.12 The differential-pressure controller can override steam flow if its controlled variable exceeds set point.

feed rate, for example, column differential pressure can rise above set point, requiring the dPC to take over manipulation of the steam valve.

Control is transferred from the steam FC to the dPC through the low-signal selector, since the selector passes whichever controller output represents the lower valve position. Most selective control systems use low-signal selectors, since the loops have been designed to fail safely on a loss of the control signal. Since there are fewer manipulated than controlled variables, all cannot be controlled at the same time. Under dP control, temperature control and hence quality regulation are sacrificed. Ideally, the overriding controller should manipulate that variable responsible for the overload so that quality control is not lost—this would be the feed rate in this example. Unfortunately, the feed rate is rarely available for manipulation, being already manipulated by another controller upstream. Even if it were, its impact on differential pressure is indirect, through the action of the temperature controller. The surest way to constrain differential pressure is therefore to override steam flow.

The positional values of controller outputs should be sent to the selector, rather than velocities or increments. Experience with velocity algorithms in selector systems has been very unsatisfactory. Noise on the dP and flow measurements passes through their controllers and is rectified by the selector, causing the selected controller to control with an offset similar to that observed with windup protection by back-calculation, but more severe.

The unselected controller must be protected from windup or constraints will be violated before switching can take place. The most effective method of windup protection for the TC is integral feedback of measured steam flow; no matter what interferes with steam flow control, the TC will not wind up. The other two controllers are protected by feeding back the selector output to both, as shown in Fig. 9.11. The unselected controller sees the current valve position fed back, which biases it in preparation to take control in the event that its deviation should go to zero. Stopping integration instead would fix the bias at some value representative of conditions existing when the last transfer took place, but this may have no relationship to the operating conditions existing when the next transfer is needed. In a steady state, both controllers will be equally biased and therefore will have identical outputs when both deviations are zero. Transfer therefore tends to take place at zero deviation. Consider, for example, the FC in a steady state at 50 percent output, with that signal biasing the dPC, whose controlled variable is below set point. Should the dPC deviation suddenly fall to zero, the two outputs will be equal, allowing transfer to the dPC output should deviation cross zero.

This response was not considered effective enough for antisurge control—there, valve motion was desired *before* set point was reached. This action can be achieved by applying the FC output as the output limit to a batch switch acting on the dPC. The batch switch would then reduce the bias of the dPC to a lower value to keep the output at that limit. The response of the dPC would then emulate that of the antisurge controller, except that its output limit would float with the load instead of being fixed at the valve limit. Back-calculation also could be used in place of the batch switch whenever the dPC output exceeded the TC output. Both these methods rectify noise as discussed earlier, however, more than the feedback arrangement in Fig. 9.11, where the two controller outputs are not forced together.

An alternative configuration to that shown in Fig. 9.11 has the selector located at the TC output. However, should the steam FC fail to control for any reason, dP control also would be lost. To maximize reliability, selectors should be placed at the lowest levels in the system.

Windup protection also has been achieved by forcing the appropriate limit of the unselected controller to track the selected output. In the case of a low-signal selector, the high limit would be made to track the selected output. A small margin must be open between the limit and the selected output to allow enough difference between the controller outputs for switching to take place. Here again, outputs are forced to track each other closely so that there is greater opportunity for noise rectification leading to offset.

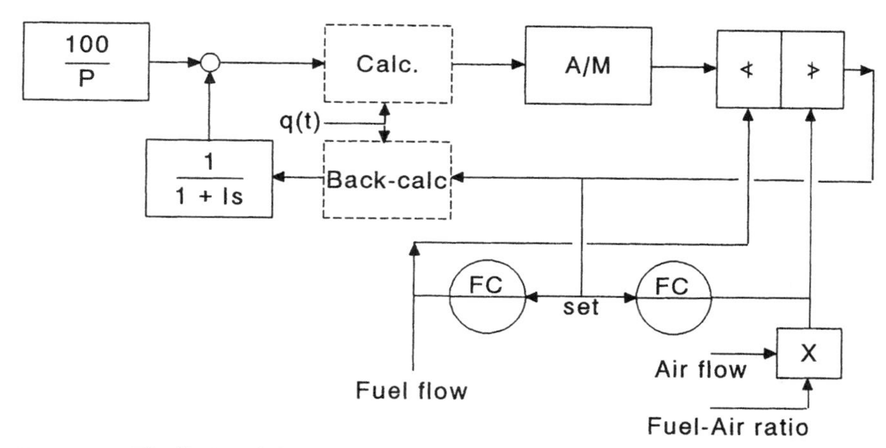

Figure 9.13 The limits of the master pressure controller are set by measured fuel and air flows to avoid a fuel-rich mixture.

As introduced under the discussion on output limits, remote setting of a limit by another controller accomplishes the same result as using a signal selector. A common application of variable limiting is the master pressure controller for a boiler, shown setting fuel and air flow controllers in cascade in Fig. 9.13. To prevent an excess of fuel from developing in case of a failure in a valve or instrument, the fuel flow measurement can be connected to the low limit of the pressure controller and the air flow measurement to the high limit. Should air flow fall for any reason, then the controller output will be driven down with it, bringing fuel flow down with air flow. Conversely, any uncontrolled increase in fuel flow will drive the low limit upward to force air flow to follow it. In normal operation, the two limits would be identical, since the two flows would be controlled at the same set point; in practice, the limits must be biased apart a percent or so to allow the pressure controller to move its output freely. For a single set point to be applied to both flow controllers, correction to the fuel-air ratio is made to the air flow measurement signal as shown.

Windup protection for the master controller is already provided by the limits, not requiring any additional consideration. If feedforward should be desired, it can be added within the master controller using the forward and back calculation blocks, which avoids interfering with the action of the limits.

Constraints are handled in large multivariable controllers by running linear or quadratic programs on the whole system at every sample interval. Discussion of these approaches is beyond the scope of this work.

Notation

b	Output bias
c	Controlled variable
d	Differential operator
D	Derivative time
D,	Distillate flow
e	Deviation
f	Function, feedback signal
F	Time constant of feedforward differentiator
I	Integral time
IAE	Integrated absolute error
IE	Integrated error
k	Ratio m/q
L	Linearity
L,	Reflux flow
m	Manipulated variable
p	Preload
P	Proportional band, percent
q	Load
$q(t)$	Dynamically compensated load signal
r	Set point
s	LaPlace operator d/dt
t	Time
v	Valve position
ß	Feedforward filter factor
σ	Standard deviation
Ω_A	Area overshoot

Subscripts

h	High limit
l	Low limit
1	Primary variables
2	Secondary variables

References

1. Shinskey, F. G., *Process Control Systems,* 3d ed., McGraw-Hill, New York, 1988, pp. 179–181.
2. *Ibid.,* pp. 470–472.
3. Ryskamp, C. J., "New Strategy Improves Dual Composition Column Control," *Hydrocarbon Proc.,* June 1980.

Index

Adaptive control, 183, 190–199
Air flow control, 269
Albert, W., 204
Aliasing error, 79, 84
Analog-digital converter, 81, 98–99
Analytical predictor, 107–109
Antisurge control, 253–254, 266
Antiwindup (*see* Windup protection)
Åstrom, K. J., 190
Auto-manual transfer, 238–243, 244, 265
Auto tuning, 183–184, 190
Automatic reset (*see* Control modes, integral; Reset)
Averaging filter, 79–80, 99
Averaging-level control, 13, 52–55, 229–231
Averaging-pressure control, 13

Back-calculation, 241–242, 257, 262–265, 268–269
Back-mixing, 37
Balance time, 242
Bang-Bang control, 221–224
BASIC code, 155–156
Batch processes, 15, 51–52, 205–207
Batch reactors, 31, 205–207, 223–224, 255–257, 260–261
Batch switch, 252–257
Bias:
  feedforward, 261–263
  output, 50–52, 54–55, 241–242
  set-point, 18, 19
Blending, 15, 37, 62
  for quality control, 8, 9
Boiler controls, 260, 269
Boilers, 4–7, 37, 105, 187
Bristol, E. H., 227
Buffering, 201–205
Burst cycling, 204
Butterworth filter, 58, 83

Capacity:
  multiple, 32–38
  non-self-regulating, 27–29
  self-regulating, 29–31
Cascade control, 199, 249, 258–262
  for reactors, 205–207, 260–261
Cement kiln, 227
Change limit, 97, 99
Characterizers, 199–205, 232–234, 244–246
Chromatographic analyzer, 77–78, 85, 88
Closed loop, 22
  gain (*see* Loop gain)
  minimum-phase, 114
  second-order, 53, 59
  time constant, 104
Combustion controls, 269
Commissioning controllers, 183–184
Compressor controls (*see* Antisurge control)
Constant of integration, 61
Constraint control, 237–269
Constraints, hard and soft, 266
Control modes:
  deadtime, 120–124
  derivative, 55–61
  double-integral, 62
  effective, 70
  filter, 54–55
  integral, 61–68
  proportional, 50–55, 242
Control valves (*see* Valves)
Controllability, 31

Controlled variable, 3
  estimated value, 102, 108, 118
Controller action, 50, 72
Controller algorithms:
  incremental and positional, 89–92
  independent and ISA, 56, 65–66
  interaction in, 69–73, 91–92
Controller gain, 50
Controller tuning (*see* Tuning)
Controllers, 3
  adaptive (*see* Adaptive control)
  Dahlin, 109–113, 120, 127–128
  digital, 88–99
  direct synthesis, 109–113
  error-squared, 53, 229–231
  FGH, 190
  Foxboro, 69, 71, 239–241
    (*See also* EXACT tuning)
  Fulscope, 72
  Honeywell, 175
  hyperactive, 26
  integral, 28, 61–65, 93–96
  internal-model, 28, 102–109, 120–121, 126–127, 131–132, 136–137
  matrix-based, 101, 116–119, 130
  model-predictive, 101, 113–119, 134–135
  Moore Products, 241
  nonlinear, 53, 211–234
  Omron, 190, 228
  on/off, 211–214, 216–220
  Pτ_d, 122–123
  PD, 55–61, 219–220, 242
  PI, 65–68, 121–123, 220
  PIτ_d, 120, 123, 127, 160
  PID, 68–75, 89–93, 121–122
    single-stage, 71–73, 257
    two-stage, 70–71
  PIDτ_d, 119–125, 127, 132, 135–138
  Powers, 188
  proportional, 49–55, 242
  proportional-lag, 55
  pulse-duration, 216–218
  sampling, 87–88, 93–96
  self-tuning, 183–207
  Smith predictor, 113–116, 120, 131, 218
  three-state, 214–216, 219–220
  time-optimal, 221–224
  Vogel-Edgar, 112–113
  Watlow, 190
  Yokogawa, 190, 228
Cost functions, 4–10
Cutler, C. R., 101
Cycle (*see* Limit cycle, Oscillation)

Dahlin, E. B., 101
Dahlin controller, 109–113, 120, 127–128
Damping (*see* Decay ratio)
Damping factor, 53, 55, 230
Dead band, 145–148
  in control valves, 145
  identification of, 146–148, 187–188
  in on/off controllers, 154, 213–215
Dead zone, 53–54, 214–216, 220
Deadbeat response, 83, 131
Deadtime, 21, 26–27
  compensation, 120–121
  controller (*see* Controllers, PIτ_d, PID τ_d)
  effective, 36–38, 86
  identification of, 143–145, 148
  margin, 43–44, 132–137, 178–179
  register, 134
Deadtime dominance, 80, 177–178
Decay ratio, 152–153, 167, 191–192
  optimum, 168, 198–199
Decoupling, 264–266
Delay ratio, 43–44
Departure trajectory, 25–26
Derivative:
  action, 55–57
    added to on/off controller, 218–220
    applied to set point, 56, 110, 175
  filter, 57–58, 129
  gain limit, 57
    effect of interaction on, 70–71, 74
    effect of sampling on, 92–93
    effect on performance, 74, 93
  inverse, 57–58
  noise sensitivity, 57
  and robustness, 178
  spikes, 85–86, 98–99, 231
  step response, 58
Deviation from set point, 50
Difference equations, 81–83, 90–92
Difference operators, 81–82, 110
Differential-pressure control, 266–268
Differentiator, 263–264
Digital controllers, 88–99
Direct synthesis controllers, 109–113
Distillation column, 34
  control, 226–227, 264–268
  dynamics, 24, 33–37, 105, 144, 187
  gain, 17

Distributed control system, 89
Distributed lag, 37–38, 161
Distribution curves, 16–18, 32
Disturbances, 3
  load, 22–25
  set-point, 3, 38
Dithering, 187
Dividers:
  as function generators, 233–234, 246
  for windup protection, 246, 262
Doss, J. E., 107
Doublet pulse, 148–149, 188–189
Drift, 186–187
Dual-mode control, 221–224
Dynamic compensation, 260, 263–264
Dynamic gain, 22–24
Dynamic matrix control (DMC), 101, 116–120
Dynamic model, 36–38, 102–103, 114, 116–118, 226

Economic criteria, 5–11
Efficiency, operating, 6–7
Einstein, Albert, 119
Endpoint control, 52, 61, 207
Energy minimization, 9–10, 14, 40–42, 223
Equal-percentage valves, 199–200, 233–234
Error in controllers, 50
Error magnitude criterion, 6–10
  best possible, 26–30, 32, 35–36
Error-squared controller, 53–54, 229–231
EXACT tuning, 184–185, 188, 191–198, 205
Exponentially weighted moving average, 15–16, 185–186

Fail-safe, 4–5, 90, 242
Failure protection, 4, 5
Feedback:
  integral (*see* Integral feedback)
  negative, 50, 71
  positive, 65, 67, 72
Feedforward control, 227–228, 260–264, 269
Fieldbus, 237–238
Filter:
  adaptive, 134–135
  averaging, 79–80
Filter (*Cont.*):
  Butterworth, 58, 83
  for chromatographic analyzers, 86
  in Dahlin controller, 110–112
  derivative, 57–58, 129, 132
  digital, 81–83
  effect on integrated error, 122
  effect on robustness, 134, 137, 178–180
  effect on tuning, 86, 162
  in internal model control, 102–104
  in level control, 54–55
  in model-predictive control, 116
  nonlinear, 230–232
  optimum, 124–125
  set-point, 58, 65, 174–177, 193
First-order lag, 29–30, 39–40
  identification of, 143–148
  negative, 31
Flow control, 184, 187, 199, 215–216
  set-point response, 172
Fouling of pH electrodes, 205
Frequency-response analysis, 142–143
Friction, 145
Fuel-air ratio control, 269
Furnace, annealing, 52, 61
Fuzzy logic, 190, 226–229

Gain:
  closed-loop, 131–132, 152–154, 194
  controller, 50
  derivative, 57–58, 70–71, 74, 92–93
  dynamic, 22–24, 159
  integral, 63–64
  open-loop, 160
    identification, 152
  proportional, 50
  steady-state, 22–24
    identification, 143–149
  valve, 199–200
Gain compensation, 199–204
Gain margin, 43–44
  in model-based systems, 131–132
Gain ratio, 43–44
Gain scheduling, 207

Hang, C. C., 190
Harmonics, 154
Heat exchangers, 37, 105, 195, 199–200
Heater:
  electric, 61, 218

Heater (*Cont.*):
  fired, 52, 61
Hold, zero- and first-order, 78–80
Horizons, 118–119
Hyperbolic characterizer, 233–234, 246
Hyperbolic cosine, 38
Hysteresis (*see* Dead band)

IAE (*see* Integrated absolute error)
IDCOM (Identification and Command), 101, 119, 135, 150, 189
Identification methods:
  closed-loop, 150–154, 189–190
    limit cycling, 153–154, 189–190
    proportional cycling, 151–153
  open-loop, 142–150, 188–189
    doublet pulse, 148–149, 188–189
    pseudorandom binary sequence (PRBS), 149–150, 189
    single pulse, 146–148, 188
    sinusoidal tests, 142–143
    step test, 143–146, 188
IE (*see* Integrated error)
Impulse function, 263–264
Incremental algorithm, 90, 240, 267
Independent algorithm, 56, 65
Instrument Society of America (ISA), 56, 65, 237
Integral controller, 28, 61–65
  sampled, 93–96
Integral feedback, 67–68, 244–248
  in batch control, 252–257
  in cascade control, 206–207, 258–261
  in feedforward control, 261–264
  in PID controllers, 69–72, 92
  in selective control, 266–269
Integral gain, 61, 97
Integral of time and absolute error (ITAE), 12
Integrated absolute error (IAE), 11, 12
  minimizing, 63, 155, 165–170
Integrated error (IE), 9–11
  in composition control, 9
  with filters, 122
  for integral controllers, 63
  for load changes:
    with analytical predictor, 108
    best possible, 27–29
    best practical, 28–32, 35
    with internal model control, 106–107
  for PI controllers, 66–67
  for PID controllers, 10–11
Integrated error (IE) (*Cont.*):
  for PID τ_d controllers, 122
  with sampling, 92
  for set-point changes, 171–176
    best possible, 38–42
    with PID controllers, 173, 256
Integrated square error (ISE), 12
Integrating process, 27–29
  with integral controller, 64–65
Interaction:
  among control loops, 196–198
  among control modes, 69–73, 92, 121, 129, 158
  among lags, 33–38
Interlocks, 4–6
Internal model control (IMC), 28, 102–109, 120–121, 126–127, 131–132, 136–137
Inventory, 9
  control, 105, 224–226
Inverse derivative, 57–58
Inverse response, 187–188
ISE (integrated square error), 12
ITAE (integral of time and absolute error), 12

Kurz, H., 204

Laboratory analyses, 88
Lag:
  digital, 81–83
  distance-velocity (*see* Deadtime)
  distributed, 37–38, 161
  first-order, 29–30, 39–40
    identification, 143–149
  interacting, 34–37, 161, 164
  multiple, 33–38, 159–161
  negative, 31
  noninteracting, 33–36, 159–161
  second-order, 32–33, 40–42, 58–61
    identification, 145, 148–149
Lag dominance, 73, 80, 84–88, 131, 144
LaPlace operator, 76
Lasher, R., 237–238
Lead, 55
  (*See also* Control modes, derivative)
Lead-lag function, 57–58
  in feedforward control, 263–264
  in internal model control, 103, 108
  as set-point filter, 175, 190
Lee, J. H., 119

Level control (*see* Liquid-level control)
Limit cycle:
  in digital systems, 96–99
  due to dead band, 146
  due to windup, 260–261
  in identification, 153–154
  with on/off control, 212–214, 217–220
Limits:
  on controlled variable, 55
  on controller output, 67, 91, 153–154, 219, 243–246
  velocity, 246–248
Liquid-level control, 28, 105, 187, 260, 264–266
  averaging, 13, 52–55, 229–231
  nonlinear, 53–54, 229–232
Load, 22–25
  dynamics, 22–24, 31, 105, 108
    effect on performance, 106–108
    effect on tuning, 162–163
  point of entry, 22–24, 26, 105
  shape, 24–25
  simulation of, 165–166
  zero, 51–52
Load response, 21–38, 191–192
  best possible, 26
  best practical, 26, 166
  with Dahlin controller, 110–112
  with internal model control, 104–109
  minimum IAE, 166–170
  ramp, 25
  sinusoidal, 25, 196–198
  step, 25–38, 155, 165–170
    overshoot, 167–170, 191
    symmetry, 167–170
Loop gain:
  closed, 152–154
  and decay ratio, 152–153
  in model-based systems, 131–132
  and overshoot, 169, 194
  variable, 199–207

Manipulated variable, 22–24
Manual control, 244–245
Manual reset, 51
Matrix-based control, 116–119, 130
Median selector, 5
Minimum-energy control, 14, 40–42, 221–223
Minimum-time control, 14, 40–42, 221–223
Mixing, 37
Model-based control, 101–138
Model error, 108–109, 114, 131–138
Model horizon, 118
Model-predictive control (MPC), 101, 113–119
Modes of control (*see* Control modes)
Modulation (*see* Pulse-duration control)
Moore, C. F., 107
Morari, M., 119
Motor, valve, 215, 220
Move supression, 118
Multicapacity processes, 34–38, 159–161
Multipliers, 262
Multivariable controllers, 119, 269

Negative feedback, 50, 71
Negative resistance, 31
Neutralization (*see* pH control)
Nichols, N. B., 50, 141, 143–145, 150–152, 164
Noise, 23
  distribution, 17, 186
  estimation, 185–187
  filtering, 57–58, 124–125, 133–135, 230–232
  and offset, 216, 252, 255, 267–268
  rectification, 90, 252, 254–255, 257, 267–268
Noise band, 186
Nonlinear compensators, 232–234
Nonlinear controllers, 211–234
  dual-mode, 221–224
  error-squared, 53–54, 229–231
  on/off, 211–214, 219–220
  PID, 229–232
  pulse-duration, 216–218
  three-piece, 53–54, 202–203, 231–232
  three-state, 215–216, 220
Nonlinear elements, 199–205
Non-self-regulation, 27–29, 103–104, 124–125
  identification of, 146–149

Objectives, 3–19
Octane control, 8
Offset, 50–52, 54
  in batch processes, 52, 61
  in cascade control, 259
  in dead zone, 214–216
  in digital control, 97–98, 267
  elimination of, 62–63

Offset (*Cont.*):
with error-squared control, 230
with on/off control, 212, 222
proportional, 50–52, 54
On/off control, 211–214, 218–219
Optimal controller settings (*see* Tuning)
Optimal switching (*see* Time-optimal control)
Oscillation:
constant-amplitude, 152
(*See also* Limit cycle)
damped, 152–153
distribution curve, 17
due to model error, 108–109, 131–138
expanding, 133, 229–230
high-frequency, 133–134
undamped, 152
(*See also* Ringing)
Output weighting, 12–14
Overrides (*see* Selective control)
Overshoot:
area, 171–172, 252
in load response, 166–169, 191, 194
of manipulated variable, 28–31, 105, 149
permanent, 52, 61
and proportional band, 194
in pulse testing, 148–149
in set-point response, 59–60, 171–177, 228
from windup, 68, 247, 253, 255–256

Pattern recognition, 191–193
Performance, 3
absolute, 42
criteria, 6, 10–18
economic, 3
of internal model controller, 106, 126–127
limits, 21–42
of model-based controllers, 125–130
of PI controllers, 63, 73–75, 127–129
of PID controllers, 73–75, 127–129
of PIDτ_d controller, 121–122, 127–130
and robustness, 42, 73–75, 136–138
Period of oscillation, 152–154
effect of decay ratio on, 152–153
effect of proportional band on, 194–198
effect of sampling on, 80
of load, 196–198
natural, 53, 59, 143, 150–154, 159, 164–165
with on/off control, 213, 218
Period of oscillation (*Cont.*):
optimum, 123, 168
of PID τ_d controller, 198–199
pH control, 187, 200–205, 240
adaptive, 201–205
batch, 61, 207
nonlinear, 202–204, 231
Phase shift:
of closed loops, 152, 159
in controllers:
with deadtime, 122–124
with derivative, 57
with integral, 63, 66
optimum, 66, 168
of sampling, 78–80
Plug-flow reactors, 37
Pneumatic controllers, 56, 71–73, 238–239, 241–243
Pneumatic transmission, 37
Pole cancellation, 102–104
in tuning PID controllers, 171–172
Positional algorithms, 89–92
Positioners (*see* Valve positioners)
Positive feedback:
in controllers, 67, 69–72, 114–115
in exothermic reactors, 31
in recycle loops, 65
Power failure, 242–243
Preload, 224, 253–256
Pressure control, 6–7, 214
averaging, 13
Pretuning, 184–190
Process gain, 22–24
Product giveaway, 8, 9
Product-quality control, 7–10
Product specifications, 8
Proportional band, 50
undamped, 152, 154, 164–165
Proportional control, 49–55
for identification, 151–153
with incremental algorithm, 91
of liquid level, 54–55
of two-capacity process, 59–60
Proportional cycling, 151–153, 190
Proportional-derivative (PD) control, 55–61
Proportional gain, 50, 70, 72
Proportional-integral (PI) control, 49, 65–68
of liquid level, 52–53, 66
Proportional-integral-derivative (PID) control, 68–75
Proportional-lag control, 55

Proportional offset, 50–52, 54
Proportional-time control (*see* Pulse-duration control)
Pseudorandom binary sequences (PRBS), 149–150, 189
Pulse-duration control, 216–218
Pulse testing, 146–150, 188–189

Quadratic system, 53, 59
Quality control, 7–10
Quality distribution, 16–19

Ramaker, B. L., 101
Ramp disturbances, 25
Rangeability, 217
Rate action (*see* Derivative, action)
Ratio control:
of fuel and air, 269
of reflux and distillate, 265
Reaction curve, 143–146
Reactors:
back-mixed, 37
batch, 31, 205–207, 223–224, 259–261
exothermic, 31, 224
plug-flow, 37
steady-state unstable, 31
temperature control in, 205–207, 223–224, 259–261
Recovery time, 167
Recovery trajectory, 25–26, 166–167
Recycle streams, 65
Redundant instrumentation, 5
Reflux, 34, 264–266
Regulation, 3
(*See also* Self-regulation)
Relay cycling, 189–190
(*See also* Limit cycle)
Repeats per minute, 61
Reset, 51, 61
(*See also* Control modes, integral)
Residence time, 9, 16, 37
Resolution, 97–99, 217–218, 224–225
Response time, 167
Richalet, J. A., 101, 134, 150
Ringing, 110–113, 124
Robustness, 21–22, 42–44
of cascade systems, 206–207
of model-based controllers, 131–138, 177, 189
and performance, 21–22, 42, 136–138, 177–180
Robustness (*Cont.*):
of PID controllers, 74–75, 137–138
plots, 42–44
tuning for, 177–180
Ryskamp, C. J., 265

Safety, 4
(*See also* Fail-safe)
Sample dominance, 83–84
Sample interval, 16, 78
effect on identification, 151
effect on integrated error, 80, 92–93
effect on performance, 79–80, 92–93, 129–130
effect on ringing, 110
effect on robustness, 95–96, 135–136, 178–179
effect on tuning, 86, 161–162
in matrix-based systems, 117, 130
selection of, 89, 130, 178–179, 185
Sampling, 77–78
aliasing error in, 79, 84
phase lag, 78–80
Sampling controller, 87–88, 93–96
Saturation of controllers, 26, 29, 86
following set-point changes, 38–42, 176–177
(*See also* Windup)
Scan period (*see* Sample interval)
Secondary lag, 58–61
effect on performance, 32–33, 40–42, 74–75, 128–129
effect on robustness, 136, 179–180
effect on tuning, 195
identification of, 144, 148–149
Secondary loop, 3
(*See also* Cascade control)
Selective control, 244, 266–269
Self-regulation, 29–31
negative, 31–32
Self-tuning, 137–138, 183–207
Servo response, 3
Set-point, 3
biasing, 18–19
optimum, 9
ramp, 62
Set-point filter, 38, 174–176, 190, 193
Set-point response, 38, 191–192
with batch controllers, 255–257
best possible, 38–42
with dual-mode control, 221–224
with filters, 174–176

Set-point response (*Cont.*):
- of flow controllers, 172
- and integrated error, 171–173
- with internal model control, 104
- with PID control, 171–177
- with PID τ_d control, 124–125, 174, 176
- with saturation, 176–177
- tuning for, 170–177
  - with interacting PID controller, 173–174
  - with pole cancellation, 171–172

Shutdown, automatic, 5–6
Signal selectors, 245, 266–269
Smith, O. J. M., 101
Smith predictor, 101, 113–116, 134–135
Solids-level control, 224–226
Stability limits, 43–44
Standard deviation, 16–19
- of noise, 17, 185–186
- of product quality, 18

Standards, 237–238
Statistical process control, 14–19
Steady-state gain, 22–24
- identification of, 143–149

Steady-state instability, 31
Step tests, 143–146
Stroking time of valves, 215, 220
Surge control (*see* Antisurge control)
Surge tank, 13, 54
- (*See also* Averaging-level control)

Symmetry:
- of cycle, 154
- of load response, 167–168

Temperature control, 7
- in distillation, 264–268
- in reactors, 31, 205–207, 255–256

Temperature controllers, 50, 256–257
Test procedures (*see* Identification)
Thermostat, 218
Three-lag process, 121–122
Three-piece nonlinear controller, 53–54, 202–203, 231–232
Three-state controller, 215–216, 220
Time constant, 29–30
- negative, 31
- primary, 29–30, 39–40
- and residence time, 9, 16, 37
- secondary, 32–33, 40–42
- thermal, 31, 205

Time delay in switching, 221–224
Time lag (*see* Lag)
Time-optimal control, 40–42, 221–224
Titration curves, 201, 203
Tracking in controllers, 248–249
- output, 248–249
- set-point, 241, 249

Trajectories, 25–26, 166–167
Transfer (*see* Auto-manual transfer)
Transmission lines, 37
Tuning, 141–180
- auto, 183–184, 190
- of Dahlin controller, 110–113, 126
- EXACT (*see* EXACT tuning)
- fine, 165–170
- of internal model controllers, 107
- for load response, 155–170
- of model-based controllers, 116
- of PID controllers, 155–180, 194–195
- of PID τ_d controller, 127–128, 155–169, 195
- for robustness, 177–180
- self, 137–138, 183–207
- of set-point filter, 174–176, 193
- for set-point response, 170–177
- of Smith predictor, 115–116, 128

Tuning rules, 155–170
- closed-loop, 164–170
  - fine tuning, 165–170, 194–195
  - initial settings, 164–165
- open-loop, 155–163
- Ziegler-Nichols, 73–74

Two-capacity processes, 58–61
- with deadtime, 74–75

Uncertainty in sampled systems, 94–96
Unstable process, 31

Valve-position control, 215
Valves:
- characteristics, 199–200, 232–234
- motors, 215, 220, 244
- positioners, 232–233
- rangeability, 96, 233–234
- response, 215, 220

Variance, 15–16
Vectors, 56, 66, 118, 132
Velocity algorithm, 90–91, 240, 267
Velocity limit, 246–248

Waveform, 24–25
Windup, 62, 68, 238, 243
in cascade systems, 90, 240
producing limit cycling, 259–261
producing overshoot, 176, 247
Windup protection, 68, 92, 238, 249–269
in cascade systems, 90, 206–207, 258–260
in digital systems, 90, 240
in feedforward systems, 260–264
in pneumatic controllers, 73, 253
in selective systems, 266–269
by stopping integration, 250–252, 254
Windup protection (*Cont.*):
by transfer to manual mode, 240
using back-calculation, 257, 262–265
using batch switch, 252–257

Zero, 103
Zero-load processes, 51–52, 61
Ziegler, J. G., 50, 141, 143–145, 150–152, 164
Ziegler-Nichols method, 73–74, 143–145, 150–152, 183, 188, 190–191

ABOUT THE AUTHOR

F. G. Shinskey was Bristol Fellow at the Foxboro Company, where he was responsible for developing new control algorithms and systems, and applying expert-systems technology to solving industrial control problems. He previously worked as a process engineer with E. I. duPont and as an instrument engineer with Olin Mathieson. Mr. Shinskey holds numerous patents on control devices, and is also the author of *Process Control Systems,* Third Edition, and *Distillation Control*, both available from McGraw-Hill, as well as *pH and pIon Control in Process and Waste Streams, Energy Conservation Through Control, Controlling Multivariable Processes,* and *Simulating Process Control Loops Using BASIC.*

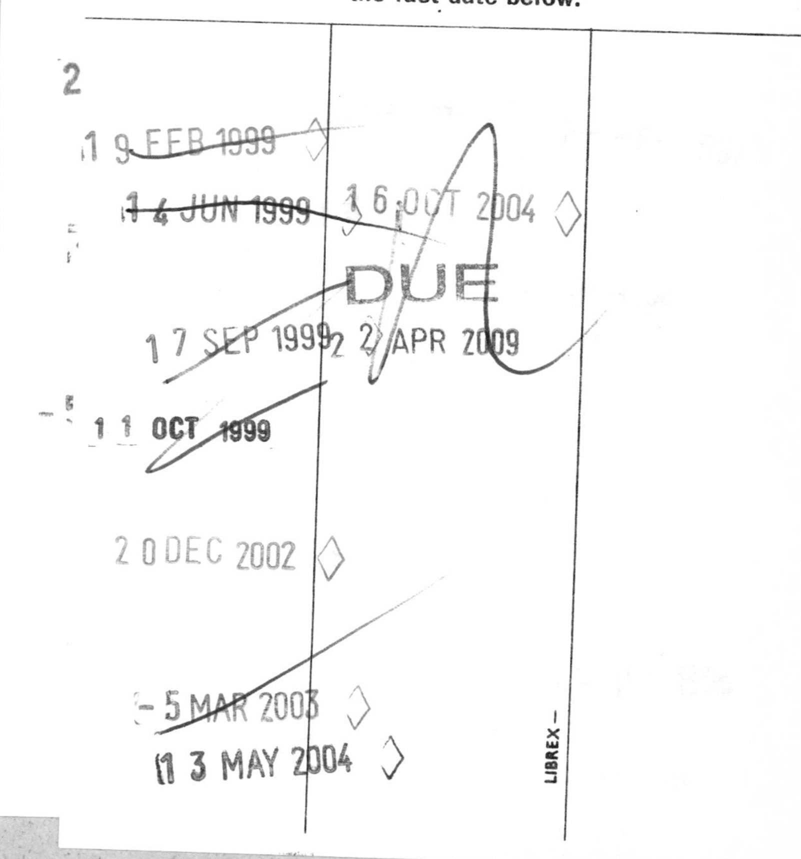